The Birth of Time

Some Books by John Gribbin

Galaxy Formation
White Holes
Timewarps
Genesis: The Origins of Man and the Universe
The Monkey Puzzle (with Jeremy Cherfas)
Spacewarps: Black Holes, White Holes, Quasars and the Universe
The Redundant Male (with Jeremy Cherfas)
In Search of Schrödinger's Cat: Quantum Physics and Reality
In Search of the Double Helix: Quantum Physics and Life
Weather (with Mary Gribbin)
In Search of the Big Bang: The Life and Death of the Universe
The Hole in the Sky
The Stuff of the Universe (with Martin Rees)
Winds of Change (with Mick Kelly)
Hothouse Earth: The Greenhouse Effect and Gaia
The Cartoon History of Time (with Kate Charlesworth)
The Matter Myth (with Paul Davies)
Stephen Hawking: A Life in Science (with Michael White)
Too Hot to Handle?: The Greenhouse Effect (with Mary Gribbin)
In Search of the Edge of Time
In the Beginning: The Birth of the Living Universe
Albert Einstein: A Life in Science (with Michael White)
Being Human (with Mary Gribbin)
Time and Space (with Mary Gribbin)
Schrödinger's Kittens and the Search for Reality
Darwin: A Life in Science (with Michael White)
Fire on Earth (with Mary Gribbin)
Watching the Weather (with Mary Gribbin)
Companion to the Cosmos
Richard Feynman: A Life in Science (with Mary Gribbin)
Q Is for Quantum: Particle Physics from A to Z
Watching the Universe
Almost Everyone's Guide to Science
In Search of SUSY
Fiction
The Sixth Winter (with Douglas Orgill)
Brother Esau (with Douglas Orgill)
Father to the Man
Ragnarok (with D. G. Compton)
Innervisions

The Birth of Time

How Astronomers Measured the Age of the Universe

John Gribbin

Yale University Press   New Haven and London

First published in Great Britain in 1999 by Weidenfeld & Nicolson

Designed by Rebecca Gibb
Set in Galliard type by Keystone Typesetting, Inc.
Printed in the United States of America by R. R. Donnelley & Sons, Harrisonburg, Virginia.

Library of Congress Cataloging-in-Publication Data
Gribbin, John R.
The birth of time : how astronomers measured the age of the universe / John Gribbin.
    p. cm.
Previously published: London : Weidenfeld & Nicolson, 1999.
Includes bibliographical references and index.
ISBN 0-300-08346-7 (c : alk. paper)
1. Cosmology.   2. Cosmochronology.   I. Title.
QB981.G756 2000
523.1 — dc21   99-0528235

A catalogue record for this book is available from the British Library.

10 9 8 7 6 5 4 3 2 1

Contents

Acknowledgements

Thanks to Simon Goodwin and Martin Hendry, who gave me a chance to do some real research once again; to Andrew Barber, for stimulating discussions about the extragalactic distance scale; and to Allan Sandage, who pointed the way.

The Birth of Time

The Age Controversy

How Wrong Is Wrong?

No child can be older than its own parent. But if you were to take some of the stories that appeared in the news media in the mid-1990s at face value, you would have thought that astronomers were stupid enough to believe that stars could be older than the Universe which gave them birth. Even the more sober reports talked of "an age crisis" (*Newsweek*, 7 November 1994), and suggested that either the astronomical understanding of how stars work must be seriously flawed, or (and, it was implied, more likely) the cherished Big Bang theory of the birth of the Universe was wrong. In that *Newsweek* report, Sharon Begley wrote:

> Don't look for a resolution to the age crisis tomorrow. As van
> den Bergh [one of the astronomical protagonists] describes
> the quest, quoting Mark Twain, "The researches of many com-
> mentators have already thrown much darkness on this subject,

and it is probable that, if they continue, we shall soon know nothing at all about it." Cosmology isn't that badly off, but there's sure to be more darkness before there is light.

It's a great quote, but she should have known better than to make predictions. Within three years of that article appearing, the age crisis, such as it was, had indeed been resolved. The reason why everyone had got so excited in 1994 was because data from the Hubble Space Telescope had been used for the first time to provide an estimate of the age of the Universe. The HST is always newsworthy, and determining the age of the Universe is indeed one of the primary functions of the telescope. But as a first result, this measurement was subject to very large potential errors, which were honestly spelled out by the astronomers involved, but were largely ignored by the news reports, which focused on the central figure quoted (giving an age of the Universe of about eight billion years) and contrasting this with estimates of the ages of the oldest stars, about fifteen billion years. It soon became clear, from entirely independent investigations, that the Universe is considerably older than this first figure from the HST implied, and that the oldest stars are significantly younger than had been thought. The reasons why this happened, and the evidence that the oldest stars are indeed younger than the Universe which gave them birth, are the themes of this book. But before we get down to the nitty gritty, I want to shed a rather different light on the nature of the "controversy" as it existed in 1995. There are degrees of being wrong, and the amount by which astronomers and cosmologists then disagreed with one another on the age question ranks pretty low on any scale of wrongness.

In case you are in any doubt about the relativity of wrong, consider a child who is asked to spell the word "car," and responds "kar." It is wrong, up to a point; but nowhere near as wrong as if the response had been "xzl." Or consider the orbit of the Earth around the Sun. Many people, asked to describe the orbit, would say that it is circular. A more accurate description is that the orbit is elliptical. But it isn't very elliptical, and the description of the orbit as a circle is not very wrong — certainly not as wrong as if you describe the orbit as being square. The amount of wrongness in astronomical ideas about the age of the Universe, as of 1994, was comparable with the amount of wrongness in saying that the spelling of "car" is "kar," or that the orbit of the Earth around the Sun is circular. But it is true that very little progress had been made in reducing that amount of wrongness for more than thirty years.

For reasons which I shall explain, by the end of the 1950s, largely as a result of the work of Edwin Hubble and his successor Allan Sandage, cosmologists knew that the maximum age of the Universe was between ten billion and twenty billion years. In everyday terms, this is a large margin of error (a factor of two). If you were buying a birthday present for a distant cousin, and couldn't remember whether she was ten years old or twenty years old, the results could be embarrassing. But to any previous generation of scientists, it would be hailed as one of the great achievements of the human intellect and the scientific method to have pinned down the age of the Universe to within a factor of two. For forty years, we have known the age of the Universe this accurately, and most astronomers have been happy to use a middle value as a rough approximation, saying that the Universe is about fifteen billion years old.

Unfortunately, starting in the mid-1970s, a split appeared in the cosmological camp. Sandage himself, and his colleagues (notably Gustav Tammann), found an increasing weight of evidence which favoured the longer end of the age range — putting it no more strongly than the accuracy of the observations justifies, a maximum age greater than fifteen billion years. Over exactly the same period, and, in many cases, using exactly the same data, a rival camp headed by Gérard de Vaucouleurs (and also championed by Sidney van den Bergh, quoted in that *Newsweek* story) favoured the shorter end of the age range — a maximum age below fifteen billion years. And in a further twist, in the 1980s the favoured cosmological models of the way the Universe expands suggested that all these numbers should be reduced by as much as (but definitely no more than) a third. Roughly speaking, this left the Sandage camp with an age of about twelve billion years, and the de Vaucouleurs camp with an age of about eight billion years (although most outsiders to the argument would have settled for ten billion years). As well as disagreeing among themselves, both camps were embarrassed by the estimated ages of the oldest stars, then still about fifteen billion years.

But not *very* embarrassed. The estimates of stellar ages were themselves uncertain, and ranged from about twelve billion years (not far from agreement with the longer estimate of the age of the Universe) to about eighteen billion years (certainly embarrassingly high if correct, but possibly wrong). And the truly remarkable thing is not the differences between all these numbers, but how close together they are. Stellar ages are calculated from the known laws of physics and observations of stars, and the numbers you get are completely independent of the way the age of the Universe is calculated, which depends on the way clusters of galaxies move apart as the Universe expands. And yet, both numbers are in the same ballpark; better still, they are both in the same part of the ballpark. If astronomers and cosmologists really didn't know what was going on, you could easily imagine that they might have come up with an "age of the Universe" of about ten thousand years, rather than ten billion years. Equally, if astrophysicists had got hold of the wrong end of their particular scientific stick, they might well have been telling us that the oldest stars are fifteen trillion years old, not fifteen billion years old. The amount of wrongness in the comparison of stellar ages with the age of the Universe is less than a factor of two — and that counts as pretty good agreement, considering what it is that is being measured.

To put this in perspective, look at the range of sizes of the objects involved in these investigations. Ignoring the fact that stars are made of tiny fundamental particles like protons and electrons, let's treat each star as a single unit. The average diameter of a star is about a billion metres. This is the size of object involved in the astronomical determination of stellar ages. At the other extreme of the cosmic entities observed by astronomers, the average size of a cluster of galaxies is about a million billion billion metres. This is the size of object involved in the cosmological determination of the age of the Universe. The difference is a span of 15 powers of 10 (15 "orders of magnitude," or 10^{15} times). But studying objects at either end of the scale gives ages which are the same as the ages derived by studying things at the other end of the scale, to within a factor of two — much less than one order of magnitude (that is, less than ten times difference). Agreement between two such important numbers, derived in such extremely different ways, to within a factor of two is, in fact, a cause for rejoicing among astronomers and cosmologists, not despair.

In my previous book on cosmology (*In Search of the Big Bang,* originally published in 1986, revised edition Penguin 1998), I didn't worry unduly about such relatively minor details as the uncertainty of a factor of two in the estimates of the age of the Universe. What mattered in telling that story was the evidence that the Big Bang really happened. But it is a sign of how much progress has been made in cosmology since the middle of the 1980s that the story of how that uncertainty in our estimates of the age of the Universe arose, and its resolution, now merits a full-length account of its own. In order to tell that new story in a reasonable space, I have given only the bare bones of the Big Bang model itself here, and if you want the full story of our search for the Big Bang you will have to find it in my earlier book, or in one of the many other popular accounts of the birth of the Universe. This book goes beyond those accounts, picking up where they leave off.

As I have said, even the minor disagreement between the two key numbers, estimates of the ages of the oldest stars and estimates of the age of the Universe, has now disappeared. It has vanished thanks to a weight of evidence gathered in since the beginning of 1995, all of it essentially pointing in the same direction. And one small piece of the jigsaw puzzle was added by a team at the University of Sussex, with which I was involved. I have often written about major scientific developments in the past, but always as an outsider, reporting on the work of other people. This time, it's personal. Although my own contribution was just one among many, I write for once as an insider, actively involved in recent years in the attempt to pin down the age of the Universe.

Like many others, the particular project I initiated was triggered by the seemingly unreasonable age determination announced by the HST team in 1994; for reasons which will become clear, and had nothing to do with the ages of stars, I was sure from the outset that this number was wrong, and determined to find a way to check it. The technique we used is disarmingly simple in principle (I had been trying to find a way to make it work, off and on, for thirty years), and is also, strictly speaking, the *only* actual measurement of distances across the Universe, and therefore of the age of the Universe itself (all the other "measurements" involve inference at some level, not pure measurement); but the technique required the observing power of the Hubble Space Telescope itself to put into practice, which is why it had never been carried out satisfactorily before 1997. As a result of this, and of the other work carried out in the past few years, I hope I can persuade you that we astronomers really do now know the age of the Universe — no mean achievement when you recall that it was only in the nineteenth century that scientists began to appreciate that there was a beginning for the Earth and Sun, let alone the cosmos at large, and began to ponder timescales dramatically longer than those favoured by contemporary theologians.

All Things Must Pass

The Discovery of Cosmic Time

When did time begin? Throughout most of human history, to most people, the question would have been meaningless. The earliest and most widespread view of time, in cultures as diverse as the Hindus, the Chinese, the civilisations of Central America, Buddhists, and even the pre-Christian Greeks, saw it in terms of cycles of birth, death, and rebirth. Like the changing cycle of the seasons, in which the Earth itself is constantly renewed, the Universe was seen as eternal, but changing in a regular rhythm. Even God is seen as being reborn, time and again, in the Buddhist and other religions.

But in the Christian religion which came to dominate the European culture from which the modern scientific investigation of the world sprang, there is only one God, and there was one unique creation event in which the Universe was born. That modern scientific investigation of the world only began in the seventeenth century, with the work of Galileo, Descartes, and Newton. Until the end of the eighteenth century, there was no conflict between the estimate of the age of the Universe calculated by theologians and the estimates made by scientists, for the simple reason that the scientists had no basis on which to make such estimates. And, rather than the vast stretch of ancient time (perhaps infinitely long) allowed by other religions, the Christian establishment taught that the world (a term synonymous in those days with the modern term *Universe*) had been created in the year 4004 B.C.

This date was not plucked at random out of the air as some wild guess by the priests, but was actually a serious attempt to relate the events described in the Bible to the world at large. It began in quite a scientific way, but the calculation became elaborated to the point of ludicrousness not long before Isaac Newton published his epic book *Philosophiae Naturalis Principia Mathematica* (generally referred to as the *Principia*), in which he laid down the principles of the scientific method which has taken us, in a little over three hundred years, from the first understanding of the orbits of the planets around the Sun to an understanding of the birth of the Universe itself, and a sound scientific determination of when that event happened.

The date for the beginning of time that Newton himself would have been taught as the Gospel truth derived initially from a calculation made by Martin Luther and his colleagues in the sixteenth century. They had based their estimate on counting back the genealogies in the Old Testament, all the way from Jesus Christ to Adam himself, and came up with a date for the Creation of 4000 B.C. This was a nice round number (what modern scientists would call an "order of magnitude" estimate), and it probably does tell us something meaningful about the timing of events described in the Bible. But in 1620, Archbishop James Ussher published his *Sacred Chronology*, which developed these ideas further. The most significant contribution made by

Ussher was to shift the timescale back by four years. Johann Kepler, the pioneering German astronomer born in 1571, had suggested that the darkening of the sky during the crucifixion of Jesus must have been caused by a solar eclipse, and by Kepler's day astronomers were able to calculate that a suitable eclipse had occurred four years earlier than the equivalent date used by Luther, the inference being that all the events in the Lutheran genealogy occurred four years earlier than he had thought — including the Creation itself.

This revision of the timescale by Ussher to set the date of the Creation as 4004 B.C. was already going far beyond the accuracy of the method being used — even if the method was a good one, and they had got the date of the crucifixion exactly right, what chance was there that all the "begats" recorded in the Bible were accurate to within four years? But the situation became even sillier in 1654, when John Lightfoot, Vice-Chancellor of the University of Cambridge, pronounced that from his study of the scriptures he had determined the final moment of the Creation, the precise moment when Adam himself was created, as 9 A.M. (Mesopotamian time) on Sunday, 26 October 4004 B.C. Isaac Newton was in his twelfth year when Lightfoot made that pronouncement, and the *Principia* would not be published until 1687. Although some may have had doubts about Lightfoot's "improvement" to the timescale, the date of 4004 B.C. for the Creation was noted in the margin of the Authorised Version of the Bible until well into the nineteenth century, when science at last became capable of mounting a serious challenge to religious orthodoxy on this point. But this is not to say that some people had not had their doubts about the biblical timescale, even before it became enshrined in the margin of the Authorised Version.

The thing that made open-minded people think that the Earth must have had a much longer history than a few thousand years was the fossil record in the rocks. Time and again over the past thousand years, different scientists were independently struck by the need for a long timescale in order to explain how the fossilised remains of diverse species got to be where they are today. The first person that we know to have puzzled over this phenomenon, and to have written his thoughts down so that they have been preserved for us to read, was the Arab scholar Abu Ali al-Hassan ibn al-Haytham, usually known to later generations of scientists by the Europeanised version of his name, as Alhazan. He was born in about A.D. 965, and died in 1038, so his productive period as a scientist covered the decades either side of A.D. 1000. He is best known for his work on optics, which remained unsurpassed for more than five hundred years (indeed, his book on optics was translated into Latin in the twelfth century, republished in Europe under the title *Opticae Thesaurus* in 1572, and was regarded as the standard text for more than a further hundred years, until Newton published his *Opticks* in 1704). But Alhazan was a wide-ranging and original thinker, who noticed the existence of fossilised remains of fish in rock strata high above sea level, in mountainous regions. He realised that the fish must have died and been covered in sediments in the ocean, and that the ocean floor had been slowly uplifted to make the mountains — a process clearly requiring a very long stretch of time, although he had no means of calculating just how long.

The standard explanation for the origin of fossils in those days was, of course, the biblical Flood. If the entire Earth really had been covered by water, including the mountain tops, then in principle there would be no difficulty in explaining how the remains of fish came to be found on mountain tops. But it wasn't just fish that showed up in the fossil record. Leonardo da Vinci (who lived from 1452 to 1519) pointed out that the fossilised remains of clams and sea-snails could be found in the mountains of Lombardy, 400 kilometres from the nearest sea (the Adriatic). There was no way that clams could travel 400 kilometres in the forty days and forty nights that the rain lasted, plus the 150 days that the waters of the Flood had, according to Scripture, covered the Earth. And there are many parts of the world where similar fossils are found much further from the present-day boundaries of the ocean.

In the seventeenth century, this kind of argument was elaborated by Niels Steensen (a Danish medical doctor who worked in Italy and wrote under the latinised version of his name, Nicolaus Steno). He noted the similarity between certain fossils and the teeth of modern sharks, and argued in a book published in 1667 the case that fossils were indeed produced by sedimentation on the sea floor, and later up-lifted by geological activity. The idea was taken up by Robert Hooke, a contemporary of Steno and one of the founders of the Royal Society. By the end of the seventeenth century, there was a serious scientific challenge to the notion of the biblical Flood as the explanation of fossils, a growing realisation that the surface of the Earth was in-volved in upheavals that could turn sea floor into mountains, and by implication a challenge to Archbishop Ussher's timescale. But there was no clear idea of the sort of timescale involved in such processes. The first step towards a scientific assessment of the age of the Earth came only towards the end of the eighteenth century, from the work of the French naturalist George-Louis Leclerc, Comte de Buffon.

Leclerc, who was born in 1707, was the son of a wealthy lawyer, and studied law himself before turning to science. He became a lead-ing naturalist, the Director of the Royal Botanical Gardens (the Jar-din du Roi) in Paris, and was made Comte de Buffon by Louis XV in 1771. Among his many interests, he was one of the first people to express clearly (in his book *Les Epoques de la Nature,* published in 1778) the idea that all the observed variety of topographical features over the surface of the Earth could be explained as a result of the slow working of processes visible today, over geological time. And, instead of invoking God as the direct, "hands on" Creator of the Earth, Buffon came up with a plausible (at the time) scientific explanation for the origin of our planet, suggesting that it had formed from a ball of molten material, torn out of the Sun by the impact of a comet. The question this raised was, how long would it have taken for this molten ball of rock to have cooled to the state it is in today?

In fact, a century before Buffon, Isaac Newton had mentioned in his *Principia* that a globe of red-hot iron as big as the Earth would take 50,000 years to cool down. But this was not taken as a serious estimate of the age of the Earth, and passed almost unnoticed alongside the much deeper scientific insights provided by the *Principia*. Buffon improved on Newton's estimate by carrying out a series of experiments with balls of iron (and other substances) of different sizes, heating them up until they glowed red and were on the point of melting, and then observing how long it took for them to cool down. Using this information, he calculated that if the Earth had indeed been formed in a molten state, it would have taken 36,000 years to cool to the point where life could exist, and a further 39,000 years to cool to its present temperature. That pushed the date of the creation of the Earth back to 75,000 years, almost twenty times further into the past than the date then enshrined in religious dogma.

Theologians of the day were unhappy about Buffon's revision of the timescale of Earth history, and the encroachment of science into what had been the theologians' preserve. It was the beginning of a debate that was to rumble on into the twentieth century, before science at last came up with a solidly based estimate of the age of the Earth that could encompass the timescales required by geology and evolution. But in the nineteenth century, even though estimates of the age of the Earth were revised upwards dramatically, compared even with Buffon's estimate, both geology and evolution were always pointing to a longer timescale than anything that could be explained by the laws of physics as they were then understood.

The next step was taken by another Frenchman, Jean Fourier, who lived from 1768 to 1830. Fourier's lasting contribution to science was the development of mathematical techniques for dealing with what are known as time-varying phenomena — Fourier analysis can be used, for example, to break down a complicated pattern of pressure variations in a sound wave into a set of simple waves, or harmonics, which can be added together to reproduce the original sound. But even many physicists and mathematicians who happily use Fourier's techniques as labour-saving devices are unaware that he developed those techniques not through any particular love of the mathematics, but because he was fascinated by the way heat flowed from a hotter object to a cooler one, and needed to develop the mathematical tools in order to describe heat flow.

Where Buffon had measured the rate at which lumps of material of different sizes cooled down, and had then extrapolated his empirical findings up to estimate the rate at which the whole Earth would cool, Fourier developed laws — mathematical equations — to describe heat flow, and then used them to calculate how long it would take for the Earth to cool. He also, crucially, made allowance for a factor that Buffon had overlooked. He realised that although the Earth is cool on the outside today, it is still hot in its interior (as the activity of volcanoes demonstrates). The temperature of molten rock, which still exists inside the Earth, is more than 6,000 degrees on the Celsius scale, and Fourier's equations could describe how heat flowed outwards from the hot interior of the planet through the layers of cooler material at the surface — layers of solid rock which act as an insulating blanket around the molten material inside the Earth, holding the heat in and ensuring that the planet takes much longer to cool down than Buffon had estimated.

And I do mean *much* longer. The number that came out of Fourier's equations was so staggering that, as far as we know, he never brought himself to write it down (or if he did, he burnt the paper he wrote it on before anyone else saw it). What he did write down, in 1820, and leave for posterity, was the formula for the age of the Earth, based on these arguments. It is easy to put the numbers into the formula and get the answer out, and Fourier must have done this for himself. But he never told anybody, because the age he came up with was beyond anyone's wildest imagination at the time — not 75 *thousand* years, but 100 *million* years. And yet, within fifty years the number that was so staggeringly large that Fourier could not bring himself to write it down in 1820 was not only widely known, but was regarded as being embarrassingly small, in the wake of the development of geological ideas about the Earth itself, and of the theory of evolution by natural selection.

Although Buffon had realised that the same physical processes that operate on Earth today could explain how the world had got into its present state, the person who first expressed the idea with full force, and who seems to have had a clear idea of just how long a period of time would be involved, was the Scot James Hutton, who was some twenty years younger than Buffon, and worked on geology in the second half of the eighteenth century. At that time, the established wisdom was that terrestrial features such as mountain ranges might indeed have been thrust upward by mighty forces, but that such events occurred catastrophically, in a very short space of time (perhaps literally overnight); it was also widely accepted that they might involve supernatural forces, and the biblical Flood was included as the classic example of such a catastrophe. By contrast, the idea that only the same natural processes that we see at work today are needed to explain how the features of the Earth have changed over time became known as uniformitarianism. In modern science, the distinction is rather blurred, because it is now accepted that what seem to be catastrophic events on any human scale (for example, the impact from space that brought an end to the era of the dinosaurs, about sixty-five million years ago) do occur on Earth from time to time. But the point to bear in mind is that on a long enough timescale, even such rare (by human standards) events are part of the natural, uniformitarian processes that have shaped the Earth. In Hutton's day, the catastrophists had to envisage all of the events that had built mountains and carved valleys, created islands and deep oceans, as having happened within the span of six thousand years — catastrophic indeed!

Hutton, who was born in 1726, studied law and medicine, but never practised either; in the early 1750s, he settled on farming as a career (his father, although primarily a merchant, owned a small estate in Berwickshire), but devoted much of his time to chemistry, while becoming increasingly intrigued by geology as a result (initially) of studying the rocky foundations of the land he farmed. In the 1760s, Hutton made a fortune out of the invention of a method for manufacturing the important industrial chemical sal ammoniac (ammonium chloride), and in 1768 he settled in Edinburgh and devoted the rest of his life (he died in 1797) to scientific pursuits.

Hutton was the first person to point out, for example, that the heat of the Earth's interior could explain, without any need for supernatural intervention, how sedimentary rocks, laid down in water, could later be fused into granites and flints. Heat from inside the Earth, he said, was also responsible for pushing up mountain ranges and twisting geological strata. And, most important of all in the present context, he realised that this would take a very long time indeed. In one striking example, Hutton came up with an analogy that used the same kind of direct, human experience that the theologians had already used in their calculations of the date of the Creation. Hutton pointed out that Roman roads, laid down in Europe two thousand years earlier, were still clearly visible, almost unmarked by erosion. Clearly, in the absence of catastrophes the time required for natural processes to have carved the face of the Earth into its present form must be enormously longer than two thousand years, and, he specifically pointed out, much, much longer than the six thousand years offered by Ussher's interpretation of Scripture. How much longer? Hutton wasn't even willing to guess. In a paper published by the Royal Society of Edinburgh in 1788, he wrote: "The result, therefore, of our present enquiry is, that we find no vestige of a beginning — no prospect of an end." He was saying that, as far as eighteenth-century science was concerned, the origin of the Earth was lost in the mists of time, and its future stretched equally incomprehensibly far into the future.

Hutton's ideas had some impact in scientific circles (and were attacked by theologians of the old school), especially after his friend John Playfair published an edited version of Hutton's writings in 1802. But Hutton's writing style was impenetrably dense (in spite of the occasional flash of a felicitous example like the one mentioned above), and his work did not make a deep impression in the world at large. Uniformitarianism, and the implied extension of the geological timescale, only really became a subject of public debate after another Scot, the geologist Charles Lyell (who was born in 1797, the year Hutton died) took up the idea and promoted it at the beginning of the 1830s.

Like Hutton, Lyell studied law — but, also like Hutton, his scientific interests soon came to dominate his life. His father was wealthy enough to support the young man, and in the 1820s he travelled widely on the continent of Europe, where he saw the evidence of the effects of the forces of nature at work first hand. He was particularly impressed by a visit to the region around Mount Etna. The fruits of Lyell's travels appeared in his three-volume work *Principles of Geology,* published between 1830 and 1833. The subtitle of the first volume of the series clearly set out his position: *Being an Attempt to Explain the Former Changes of the Earth's Surface by Reference to Causes Now in Operation.* Unlike Hutton, Lyell wrote clearly and accessibly, opening up these ideas for any educated person of the time. Indeed, the work is still just as accessible, still worth reading, and is available in a Penguin Classic edition.

Lyell's book made a particularly striking impression on the young Charles Darwin. Darwin had been born in 1809, and at the time he set out on his famous voyage in the *Beagle,* at the end of 1831, he regarded himself, in scientific terms, as primarily a geologist. He took the first volume of Lyell's masterwork with him on the voyage; the second volume caught up with him during the ship's circumnavigation of the globe; and the third volume was waiting for him when he returned to England in 1836. He later wrote that the book "altered the whole tone of one's mind . . . when seeing a thing never seen by Lyell, one yet saw it partially through his eyes." And, in the most telling phrase of all, in the context of his theory of evolution by natural selection, Darwin said that Lyell had given him "the gift of time." For, of course, the theory of natural selection also explains how great changes have been brought about by very slow, uniformitarian processes operating over enormous amounts of time. Evolution by natural selection can only explain how the variety of forms of life on Earth evolved from a common ancestor if there has been a huge span of time during which evolution could do its work. After the *Origin of Species* was published in 1859, both biology and geology were telling scientists that the Earth must be very ancient indeed, with "no vestige of a beginning"; and, by implication, the Sun must be at least as old as the Earth, or life could not have existed and evolved on Earth over the requisite time span. This threw the biologists and geologists into direct conflict with the physicists, and in particular with the greatest physicist of the time, Lord Kelvin.

The problem was not that the physicists didn't know what was going on. Quite the reverse. By the middle of the nineteenth century they understood the laws of physics well enough to be able to say with absolute confidence that, according to the known laws of physics, there was absolutely no way that the Sun could have been shining for as long as Darwin and the geologists required.

Lord Kelvin started life (he was born in 1824) as plain William Thomson, and he was another of the Scots that played a large part in the development of British science. As well as being a great physicist, Thomson was very practically minded, and in the great tradition of the Victorian entrepreneurs he applied his talents not just to science but to engineering, making a fortune from his patents and being the brains behind the first successful transatlantic telegraph cable (as profound a development in the 1860s as the development of the communications satellite was to be in the 1960s). It was for his services to industry, adding to the wealth of Britain, not his scientific work, that he was first knighted (in 1866) and then ennobled, in 1892, as the first Baron Kelvin of Largs. Although most of his great scientific work was behind him by the time he became a peer, he is usually referred to even in scientific circles today simply as Lord Kelvin, and the absolute scale of temperature, which he devised from the fundamental principles of thermodynamics, is now known as the Kelvin scale, not the Thomson scale. Zero on the Kelvin scale is at -273 degrees on the Celsius scale, but the size of each degree on the Kelvin scale is the same as on the Celsius scale.

Kelvin was the towering figure in physics in Britain in the second half of the nineteenth century, and almost as dominant in the context of European science. He graduated with high honours from the University of Cambridge in 1845, and a year later, at the age of twenty-two, became professor of natural philosophy (the old name for what we now call physics) at the University of Glasgow. He held the post for fifty-three years, until he retired in 1899. Among his many achievements in science, Kelvin laid the foundations of thermodynamics (formulating the famous Second Law of Thermodynamics, which says that heat cannot flow unaided from a cooler object to a hotter one, in 1851), and helped to develop the theory of the electromagnetic field. It was through his study of thermodynamics that he was led to ponder the question of the ages of the Earth and the Sun.

The most important thing that thermodynamics teaches us is that nothing lasts forever. All things must pass, and everything wears out. This led Kelvin to exactly the opposite conclusion from that drawn by Hutton concerning the history of the Earth. In 1852 he wrote: "Within a finite period of time past the earth must have been, and within a finite period of time to come the earth must again be, unfit for the habitation of man as at present constituted, unless operations have been, or are to be performed which are impossible under the laws to which the known operations going on at the present in the material world are subject." Of course, there is not really any conflict between Kelvin's assertion that the age of the Earth is finite and Hutton's assertion that there was no vestige of a beginning in the geological record. We now know that Kelvin's finite age is so enormously long that no vestige of the beginning could, indeed, have been detected in the eighteenth century, even though we can perceive it quite clearly today. But there are two points worth picking up concerning Kelvin's comment, and his continuing contribution to the debate over the next half-century (he died in 1907). The first is that Kelvin was working rigorously within the laws of physics known in his day; the second is that because of his huge prestige, the view that the Earth had a relatively short lifetime, certainly much shorter than that required by the geologists and evolutionists, held sway right into the twentieth century.

Partly under the stimulus of pressure from, first, the geologists and later the evolutionists, Kelvin refined his thermodynamic arguments in successive stages through the second half of the nineteenth century. Some of his ideas built from the work of his British contemporary John Waterston, and a lot of Kelvin's thinking about the way the Sun might gain its energy was duplicated by the German physicist Hermann von Helmholtz, leading to one of those bitter debates about priority that all too often plague science. But there is no need to follow every step in Kelvin's development of his ideas about the Sun, and today we can readily agree that Helmholtz had the same idea independently, so that the timescale for the life of the Sun (or, indeed, any star) that comes out of the calculation is often called the Kelvin–Helmholtz timescale (in Germany, of course, it is known as the Helmholtz–Kelvin timescale). The complete version of the idea was presented in a lecture by Kelvin at the Royal Institution, in London, in 1887. It is impeccable science, and it goes like this.

The Sun is a very large ball of gas, with a mass roughly 330,000 times the mass of the Earth, and a diameter roughly 109 times the diameter of the Earth. It ought to be shrinking, under its own weight; but it is held up by the pressure associated with the heat of its interior. But that heat has to come from somewhere — the laws of thermodynamics spelled out, more clearly to Kelvin's generation than ever before, that there must be a source of energy to keep the Sun shining. The major sources of energy known in Kelvin's day were chemical sources, and the industrial revolution in Britain was being fuelled by the combustion of coal. But it was easy to calculate that if the Sun were entirely made of coal, burning in pure oxygen, it could maintain its output of energy for only a few thousand years. Updating the argument to express it in terms of the fuel that powers the modern industrialised world, if the Sun were made entirely of gasoline burning in pure oxygen, it could maintain its present heat for only about thirty thousand years.

The insight which Kelvin and Helmholtz thought up independently was that there is actually another source of energy, other than chemical energy, which the Sun can draw on — gravity. When an object falls in a gravitational field, its gets accelerated, picking up kinetic energy (energy of motion). If it is then brought to an abrupt halt by hitting the ground, this kinetic energy is turned into heat, as it is dissipated as thermal motion among the atoms and molecules that make up the object (and the atoms and molecules of the object it hits). In the early stages of the development of the idea, Kelvin considered how much heat might be released by allowing meteors, comets, or even whole planets to collide with the Sun. But then he realised that this is not necessary. The greatest source of gravitational energy, as far as the Sun is concerned, is the most massive object in the Solar System, the Sun itself.

It is one of the insights from thermodynamics that heat is associated with atoms and molecules moving about and colliding with one another — the faster they move, the hotter the object is. If you imagine all the material that now makes up the Sun dispersed into a thin cloud in space, then falling together under the influence of gravity to make the Sun, it is easy to see how gravitational energy will be converted into heat as all the atoms and molecules move faster and faster, and collide with one another. Indeed, this is still the way that astronomers believe stars form and get hot in the first place. The additional insight from Kelvin and Helmholtz is that even in its present state, as a relatively compact, hot ball of gas, the Sun can draw on its remaining reserves of gravitational energy and turn them into heat by shrinking slowly. Shrinking means that all the particles in the Sun move closer to the centre, falling in its gravitational field and gaining kinetic energy, so that they jostle one another more vigorously, getting hot. If the Sun were shrinking at a rate of only 50 metres a year, Kelvin calculated, it would release enough energy to explain its observed brightness. This amount of shrinking was far too small to be detected by astronomers in the nineteenth century, so there was no obvious reason to reject the idea. It extended the timescale available for geology and evolution enormously — but not enormously enough. In round terms, the Kelvin–Helmholtz timescale says that a star like the Sun must fizzle out in about twenty million years. And this was still far too short to satisfy the needs of geology and evolution. The more clearly Kelvin expressed his argument, and the more accurately he refined his calculations, the more obvious it became that there really was a conflict.

In 1892, the year he received his peerage, Kelvin returned to the remark he had made in 1852, and updated it: "Within a finite period of time past the earth must have been, and within a finite period of time to come must again be, unfit for the habitation of man as at present constituted, unless operations have been and are to be performed which are impossible under the laws governing the known operations going on at present in the material world." And by 1897 he had set the upper limit on the lifetime of the Sun as twenty-four million years. But, in exactly the decade that Kelvin was reaching these conclusions, based on impeccable application of the known laws of physics, other scientists were realising that what he referred to as "the laws governing the known operations going on at present in the material world" were not the whole story. The discovery of radioactivity revealed the existence of previously unknown laws of physics, and previously unknown sources of energy, which would soon resolve the conflict between the timescales of geology and evolution and the timescale of the Sun.

The 1890s were exciting times for physics. The term *revolution* is over-used as much in science as in other walks of life — but the events following the discovery of X-rays in 1895 were as revolutionary as anything that has ever happened in science.

X-rays were discovered by the German physicist Wilhelm Röntgen in 1895, and the discovery was announced on 1 January 1896. Röntgen had been studying what were then called cathode rays (we now know them to be streams of electrons), produced from the negatively charged plate of an electric discharge tube (a "vacuum tube," or cathode ray tube, not unlike the picture tube in a modern TV set). He discovered, by chance, that the cathode rays striking the glass wall of the tube produced a secondary form of radiation, which made a detector screen nearby, painted with barium platinocyanide, glow when the tube was switched on. Although this previously unknown form of radiation was initially called "Röntgen radiation," it soon became known as X-radiation, after the familiar mathematical symbol for the unknown quantity.

The discovery of X-radiation encouraged other physicists to search for "new" forms of radiation, and the most spectacularly successful of these seekers was Henri Becquerel, working in Paris. Because Röntgen had discovered that X-rays come from a bright spot on the wall of the vacuum tube, where the cathode rays made the material of the glass fluoresce, Becquerel looked for similar kinds of activity associated with phosphorescent salts (salts that glow in the dark), including some uranium salts. Phosphorescent material is usually "charged up" by being exposed to sunlight, and glows for a while afterwards, before the glow fades and the material has to be recharged by a further dose of sunlight. Becquerel soon found that some of his phosphorescent salts didn't just produce a visible glow in the dark, but also produced yet another kind of radiation. This radiation could escape and fog a photographic plate nearby, even when the plate was wrapped in thick black paper. This was exciting enough in itself. But at the end of February 1896 Becquerel made a sensational discovery.

In his latest series of experiments, Becquerel had prepared a photographic plate, wrapped in thick black paper so no light could penetrate, and a piece of copper in the shape of a cross (he had already found that the new radiation could not penetrate metal). The copper cross sat on top of the wrapped photographic plate, and a dish of uranium salts sat on top of the copper. Becquerel planned to expose the salts to sunlight and see if the resulting activity of the salts produced enough radiation to make an imprint of the outline of the copper cross (a kind of radiation shadow) on the photographic plate. Because the skies over Paris were overcast for several days, Becquerel left the prepared experiment in a cupboard, ready and waiting. Then, perhaps because he had got bored, he developed the photographic plate anyway, even though the experiment had not been exposed to sunlight. It showed a clear image of the copper cross. Becquerel had not only discovered a new form of radiation (soon to be called radioactivity); he had discovered a new form of energy, because the activity of the salts clearly did not require an input of energy from the Sun (unlike normal phosphorescence), nor was there any "man-made" input of energy to the system, like the electricity which drove the cathode rays to make the X-rays in Röntgen's experiment. It looked as if uranium salts could sit quietly radiating energy out into the world at large from no visible source, in seeming contradiction to one of the most cherished laws of science, the law of conservation of energy.

Becquerel's discovery was taken up and carried forward by the husband and wife team of Marie and Pierre Curie, also working in Paris. It was Marie Curie who introduced the term *radioactive substance,* in a paper published in 1898. The team showed that the amount of radioactivity in a sample of salts containing uranium depended on the amount of uranium in the sample (so it was clear that the radioactivity came from uranium itself), and they identified two previously unknown radioactive elements (that is, previously unknown elements, not just known elements that were not known to be radioactive), polonium and radium. The key implication of this work is that radioactivity is a property of the individual atoms of an element — it is not something to do with the chemistry of uranium salts, or any other compound. And this was all going on at the same time that, over in Cambridge, J. J. Thomson (no relation to Lord Kelvin) was discovering that cathode rays are actually tiny charged particles, the particles we now call electrons, which had somehow been chipped away from atoms, which had previously been regarded as the indestructible and unchanging building blocks of matter.

The person who put all of the pieces of the puzzle together, coming up with a new timescale for the Earth and pointing the way towards a new energy source for the Sun, was the New Zealand–born physicist Ernest Rutherford, who had been born in 1871 (he lived until 1937). Rutherford worked with Thomson in Cambridge in the 1890s, before moving to McGill University, in Montreal, in 1898; he stayed there until 1907, when he took up a post at the University of Manchester, in

England. He moved again, to become Director of the Cavendish Laboratory in Cambridge, in 1919, and stayed there for the rest of his career.

The great thing about radioactivity is that it gives you both a timescale and an energy source, in one package. Rutherford showed that the radiation Becquerel had discovered was actually a mixture of two kinds of radiation, which he called alpha rays and beta rays. It has since been established that beta rays are fast-moving electrons, like cathode rays but carrying much more energy. Rutherford himself showed that alpha rays are a stream of particles, that each alpha particle has the same mass as four hydrogen atoms, and that each alpha particle carries two units of positive charge. He concluded, correctly, that an alpha particle is identical to a helium atom that has lost two units of negative electric charge — that it has lost two electrons. This was a significant step forward, less than ten years after the identification of the electron itself as a component of atoms.

Jumping ahead in our story a little, to Rutherford's work in Manchester, it was also a team under Rutherford's direction that discovered the basic structure of the atom. Hans Geiger and Ernest Marsden fired alpha particles (produced by natural radioactivity) towards thin sheets of gold foil, and were surprised to discover that although most of the alpha particles passed right through the foil as if it were not there, just occasionally one of the alpha particles bounced back as if it had struck something solid. Rutherford interpreted these results as indicating that every atom consists of a very compact core of positively charged material (which he called the nucleus) surrounded by a tenuous cloud of negatively charged electrons. An alpha particle can brush through the electron cloud as if it were not there, like a cannonball whizzing through a fog bank. But, just occasionally, an alpha particle (which itself has positive charge) will meet an atomic nucleus more or less head on, and be deflected by electrical repulsion (as if the cannonball whizzing through the fog bank hits a solid object concealed by the fog, and bounces off). To put this in perspective, the largest atom is just 0.0000005 millimetres (that is, 5×10^{-7} mm) across; within any atom, the size of the nucleus compared with the size of the electron cloud that makes up the bulk of the atom is in the same proportions as a grain of sand to the volume of the Albert Hall.

Armed with this image of an atom, we can go back to the story of radioactivity. A hydrogen atom is regarded as being made up of a single proton (relatively massive, and carrying one unit of positive charge) and a single electron (with only one two-thousandth the mass of a proton, and carrying one unit of negative charge); a helium atom has a nucleus containing two protons and two neutrons (electrically neutral particles almost identical in mass to a proton) with two electrons outside the nucleus. An alpha particle is exactly the same as a helium nucleus that has no electrons associated with it. And either electrons (beta rays) or helium nuclei (alpha rays) can be ejected from the nuclei of radioactive atoms in the process of radioactive decay. Working with Frederick Soddy, in Canada, Rutherford explained that radioactivity is associated with the disintegration of atoms (thanks to his later work, we would now say the disintegration of nuclei), when atoms of the radioactive element are converted into atoms of another element. This immediately tells us that the source of energy associated with radioactivity is, after all, finite, and does not violate the law of conservation of energy. Radioactivity involves a rearrangement of the nuclei of atoms into more stable, lower energy states, with the "spare" energy being released along the way. This is exactly equivalent to the way energy is released by chemical reactions (for example, by burning) when atoms are rearranged into lower energy states and the spare energy is released (in this case, as heat and light) along the way. Once all the original radioactive atoms in a sample have disintegrated in this way, the radioactivity, and the release of energy, will stop — but it may be a very long time before this happens.

Rutherford also discovered that whatever amount of radioactive material (in the form of a pure radioactive element, such as radium or uranium) that you start out with, half of the atoms in the sample will decay in this way in a certain amount of time, now called the "half-life" of the element. He didn't even have to wait for many years to measure the half-lives of interesting substances, because this kind of law can be extrapolated from measuring the way in which the rate at which a sample decays in the laboratory changes over a much shorter period of time (this was, of course, one of the first things Rutherford looked at, to see if the radioactivity of his samples was decreasing as time passed, as it must do if the law of conservation of energy holds).

In a sample of radium, for example, after 1,602 years just half of the atoms will have decayed into atoms of the gas radon, as alpha and beta particles are ejected from the original radium nuclei. In the next 1,602 years, half of the rest of the sample (one quarter of the original atoms) will decay in this way, and so on. This is one of the strange features of the rules of quantum physics, which govern the behaviour of things on the scale of atoms and below; it was the discovery of this kind of behaviour which led Albert Einstein to comment in despair, "I cannot believe that God plays dice." But all the evidence is that Einstein was wrong; in effect, an individual atom (strictly speaking, an individual nucleus) does "play dice," as if at some randomly chosen instant during each half-life each atom (nucleus) rolled a single die, and decayed if the number that came up was odd, but didn't decay if the number that came up was even. An individual nucleus may decay in the next second, or not for thousands of years, and there is no way to tell in advance what it will do. But over a large enough collection of nuclei, the overall behaviour of the sample becomes very regular and predictable. Don't worry about the quantum physics; all that matters for our present story is that, as Rutherford realised, this provides a clock which can be used to measure the age of the Earth.

Provided you know how many radioactive atoms you started out with in a sample of rock, all you have to do is to measure how many are left (by measuring the strength of the radioactivity of the sample) to know exactly how many half-lives have elapsed since the rock was formed. But how do you know how much radioactivity there was in the rock in the first place? The first handle on the problem is to find a radioactive decay process which produces a stable product that would not otherwise be present at all in the samples being studied. Then, simply by measuring how much of this "daughter" product is present you know how much of the radioactive "parent" has decayed already.

Rutherford himself first tackled the problem, in 1905, by measuring traces of helium trapped inside rocks that contain uranium compounds. The helium could only have been produced by alpha particles from the decay of the uranium, with each alpha particle latching on to two electrons to become an atom of helium. This gave Rutherford and his colleague Bertram Boltwood (an American chemist chiefly based at Yale, who visited Manchester in 1909–10) an estimate of 500 million years for the ages of the relevant rocks. Since any helium that had been present when the rocks were in a molten state would have escaped, and some may have seeped away through cracks in the rock, this was the minimum time since those rocks were laid down — very much a minimum age for the Earth. Yet it was twenty times longer than the maximum timescale that Kelvin had calculated for the Sun less than ten years before — and this was just the beginning.

Boltwood's key contribution was to take the technique a stage further, looking at all of the products of uranium decay, not just helium. He realised that the ultimate stable product that uranium is transformed into by decay is lead, with radium as an unstable intermediate product. With the decay rates (half-lives) of both uranium and radium known, it was possible in principle to determine the ages of rocks by measuring the amounts of all these substances in them today, assuming that no lead was present at the start. The practical side of this work was far from easy — it involved measuring accurately traces of radium amounting to only 380 parts per billion, in various samples of rocks. But by the end of the first decade of the twentieth century it was giving ages for various rock samples in the range from 400 million years to more than two billion years, albeit with some uncertainty in the estimates.

Both Rutherford and Boltwood went on to other work, but the torch was taken up by Arthur Holmes, then working at Imperial College, in London. Holmes dated many rock samples using the uranium–lead technique, and by 1913, he had come up with an age of 1.64 billion years for the oldest of these samples, with relatively small experimental errors. It was Holmes who made the whole business of radiometric dating, as it became known, respectable. He was the first person to use radioactive dating (the term is used synonymously with radiometric dating) to determine the ages of fossils, putting absolute dates into the fossil record for the first time, and over the years that followed he extended the technique by taking on board new ideas and discoveries, most notably the fact that many elements come in different varieties, called isotopes.

All isotopes of an element have the same chemical properties, because each atom has the same number of protons in its nucleus, and therefore the same number of electrons in the cloud around the nucleus. As far as chemistry is concerned, almost all that matters is the number of electrons in the cloud, which is the visible face that an atom shows to other atoms. But different isotopes of the same element have different numbers of neutrons in their nuclei, so they have different masses. The total number of neutrons in the nucleus affects the stability of the nucleus. For example, uranium actually comes in different varieties, the most relevant here being U-238 and U-235. Each uranium atom has 92 protons in its nucleus, but each nucleus of U-238 contains, in addition, 146 neutrons, while each nucleus of U-235 contains 143 neutrons along with its 92 protons. As a result, U-238 (which makes up about 99 percent of all naturally occurring uranium on Earth) has a half-life of 4.51 billion years, while U-235 (which makes up about 0.7 percent of all naturally occurring uranium on Earth) has a half-life of just 713 million years. There are other, rarer isotopes of uranium, but they need not concern us here. What matters, without going into details, is that once scientists understood the nature of isotopes, and had the techniques required to measure the relative abundances of different radioactive isotopes and their daughter products in rock samples, the whole radiometric dating business became that much more accurate.

By 1921, a debate at the annual meeting of the British Association for the Advancement of Science showed that there was a new consensus. Geologists, biologists, zoologists and now the physicists as well all agreed that the Earth must be a few billion years old, and they all agreed that the radiometric dating technique provided the best guide to its age. The final seal of approval came in 1926, in the form of a report from the National Research Council of the US National Academy of Sciences, which endorsed the technique. Since the 1920s, further refinements of the technique (and the discovery of particularly ancient rocks at some sites on Earth) have pushed back the radiometrically determined ages of the oldest known rocks still further. Holmes

himself continued to work on the technique (alongside other research) until the end of the 1950s (he died in 1965, at the age of
seventy-five), and the current estimate for the ages of the oldest rocks
on Earth is 3.8 billion years. Even that, though, is not the end of the
story — material from meteorites, pieces of rocky debris that fall to
Earth from space, has been dated in the same way, and the oldest of
these pieces of cosmic debris have ages around 4.5 billion years. Since
meteorites are thought to be samples of the rocky material that was
left over from the formation of the planets when the Solar System
was born, this is now the best measurement we have of the age of the
Solar System, and, by implication, the age of the Sun. Not merely
twenty times Kelvin's estimate, based on the accurate application of
the known laws of nineteenth-century physics, but *two hundred* times
Kelvin's estimate. The reason for the discrepancy, of course, is that
there are laws of physics which were not known to nineteenth-century
science.

The first clue comes from radioactivity itself. Radioactive decay
releases energy that has been stored in the nuclei of atoms. In the case
of long-lived isotopes such as U-238, the energy may have been stored
in this way for billions of years, since the uranium was manufactured.
(How did the energy get in there in the first place? It was put there by
the explosion of a dying star, as I shall shortly explain). What Buffon
and Fourier and their contemporaries could not know is that the
Earth has not simply cooled down into its present state from a molten
glob of material, but its internal heat is maintained by the energy
released in radioactive decay still going on in its interior. This pushes
back estimates of the "cooling age" of the Earth into the same region
of time, billions of years, indicated by the radiometric dating. And it
was very quickly apparent to Rutherford's generation of physicists

that some form of radioactive energy source might keep the Sun shining for a comparably long interval.

When Lord Kelvin remarked, as he often did late in his career, that the only way to provide a timescale for the Sun longer than a few tens of millions of years would be to invoke unknown sources of energy and new laws of physics, it is clear from the context of these remarks that he meant them to be taken as ridiculing such notions, not something to be taken seriously. Right at the end of the nineteenth century, though, the American geologist Thomas Chamberlin, acutely aware of the new discoveries made by Becquerel and the Curies, made a much more prescient comment, in the journal *Science* (volume 10, page 11):

Is present knowledge relative to the behavior of matter under such extraordinary conditions as obtained in the interior of the sun sufficiently exhaustive to warrant the assertion that no unrecognised sources of heat reside there? What the internal constitution of the atoms may be is yet open to question. It is not improbable that they are complex organisations and seats of enormous energies. Certainly no careful chemist would affirm that the atoms are really elementary or that there may not be locked up in them energies of the first order of magnitude. No cautious chemist would . . . affirm or deny that the extraordinary conditions which reside at the center of the sun may not set free a portion of this energy.

But just how much energy do we have to unlock from these "seats of enormous energies" to keep the Sun shining? One of the most striking analogies was made by the physicist George Gamow, in his book *A Star Called the Sun,* published in the early 1960s. If an electric coffee percolator is advertised as being so effective that it produces heat at the same rate as heat is produced (on average) over the entire volume of the Sun, he asked, how long would you wait for the pot to boil the water to make the coffee? The surprising answer to Gamow's question is that even if the pot were perfectly insulated so that no heat could escape while you were waiting, it would take more than a year (the time does not depend on the size of the coffee pot) for the water to boil.

The key to the puzzle is that *on average* each gram of the mass of the Sun produces very little heat. Astronomical measurements show that 8.8×10^{25} calories of heat energy cross the Sun's surface each second. But the mass of the Sun is 2×10^{33} grams. So *on average* each gram of material inside the Sun generates a mere 4.4×10^{-8} calories of heat per second. This isn't just low by the standards of heat generation in the average coffee percolator — it is much less than the rate at which heat is generated in your body through the chemical processes associated with human metabolism.

If the whole Sun were just slightly radioactive, it could produce the kind of energy that we see emerging from it in the form of heat and light. In 1903, Pierre Curie and his colleague Albert Laborde actually measured the amount of heat released by a gram of radium, and found that it produced enough energy in one hour to raise the temperature of 1.3 grams of water from 0°C to its boiling point. Radium generates enough heat to melt its own weight of ice in an hour — every hour. In July that year, the English astronomer William Wilson pointed out that in that case, if there were just 3.6 grams of radium distributed in each cubic metre of the Sun's volume it would generate enough heat to explain all of the energy being radiated from the Sun's surface today. It was only later appreciated, as we shall see, that the "enormous energies" referred to by Chamberlin are only unlocked in a tiny region at the heart of the Sun, where they produce all of the heat required to sustain the vast bulk of material above them.

The important point, though, is that radioactivity clearly provided a potential source of energy sufficient to explain the energy output of the Sun. In 1903, nobody knew where the energy released by radium (and other radioactive substances) was coming from; but in 1905, another hint at the origin of the energy released in powering both the Sun and radioactive decay came when Albert Einstein published his special theory of relativity, which led to the most famous equation in science, $E = mc^2$, relating energy and mass (or rather, spelling out that mass is a form of energy). This is the ultimate source of energy in radioactive decays, where careful measurements of the weights of all the daughter products involved in such processes have now confirmed that the total weight of all the products is always a little less than the weight of the initial radioactive nucleus — the "lost" mass has been converted directly into energy, in line with Einstein's equation.

Even without knowing how a star like the Sun might do the trick of converting mass into energy, you can use Einstein's equation to calculate how much mass has to be used up in this way every second to keep the Sun shining. Overall, about 5 million tonnes of mass have to be converted into pure energy each second to keep the Sun shining. This sounds enormous, and it is, by everyday standards — roughly the equivalent of turning five million large elephants into pure energy every second. But the Sun is so big that it scarcely notices this mass loss. If it has indeed been shining for 4.5 billion years, as the radiometric dating of meteorite samples implies, and if it has been losing mass at this furious rate for all that time, then its overall mass has only diminished by about 4 percent since the Solar System formed.

By 1913, Rutherford was commenting that "at the enormous temperatures of the sun, it appears possible that a process of transformation may take place in ordinary elements analogous to that observed in the well-known radio-elements," and added, "the time during which the sun may continue to emit heat at the present rate may be much longer than the value computed from ordinary dynamical data [the Kelvin–Helmholtz timescale]."

So, by the beginning of the third decade of the twentieth century the great age debate had moved firmly off the surface of the Earth and out into space. The scientific evidence that the Earth was a few billion years old was compelling, and there were clear hints from the special theory of relativity and from the existence of radioactive elements on Earth that there was a source of energy which could keep the Sun and stars shining for at least that long. But how did they do the trick? And just how old were the oldest stars?

2

Age Limits

The Oldest Things in the Universe

Before the 1920s, nobody knew how stars work. The idea that the Sun might be generating heat by contraction still lingered, in spite of the timescale problem, and in spite of the evidence from studies of radioactivity that the nuclei of atoms contained a store of energy unknown to Kelvin and his predecessors. This was simply because nobody had yet come up with an explanation of how nuclear energy, as it is now called, could be released steadily inside stars. The person who pointed astronomers in the right direction, and virtually invented the scientific discipline of astrophysics, was the British astronomer and physicist Arthur Eddington.

Eddington had been born in 1882, and graduated from the University of Cambridge in 1905, the year that Einstein published his special theory of relativity, so he was a member of the first generation of researchers to come across the ideas of relativity theory (including $E = mc^2$) at the very beginning of their career. Among his many other achievements (he became Plumian Professor of Astronomy and Experimental Philosophy at the University of Cambridge in 1912, at the age of twenty-nine, and Director of the Cambridge Observatories in 1914), Eddington was the academic Secretary of the Royal Astronomical Society in 1915, when Einstein completed his general theory of relativity. Although this was in the middle of World War One, and there were no direct scientific links between Britain and Germany, Einstein sent copies of his papers to Willem de Sitter, in neutral Holland, and de Sitter passed them on to Eddington, in his official capacity at the Royal Astronomical Society. Eddington was in the perfect position to spread the news of Einstein's achievement in the English-speaking world, and quickly became the leading authority on the new theory outside Germany. Soon after hostilities ended, it was Eddington who led the eclipse expedition that, in 1919, successfully measured the way light is bent by the Sun, confirming a prediction made by Einstein's theory. Einstein immediately became a global figure, the archetypal image of the scientific genius; in Britain, Eddington's prestige was almost as great as that of Einstein, and his influence was immense. His lasting achievement was to apply the laws of physics to the conditions that operate inside stars, explaining their overall appearance in terms of the known laws relating the temperature, pressure, density, and so on inside them.

It took time — decades — before all the details were fully worked out. But Eddington nailed his colours to the mast in a talk he gave in 1920, at the annual meeting of the British Association for the Advancement of Science, held that year in Cardiff:

> Only the inertia of tradition keeps the contraction hypothesis
> alive — or rather, not alive, but an unburied corpse. But if we
> decide to inter the corpse, let us freely recognise the position in

which we are left. A star is drawing on some vast reservoir of energy by means unknown to us. This reservoir can scarcely be other than the sub-atomic energy which, it is known, exists abundantly in all matter; we sometimes dream that man will one day learn to release it and use it for his service. The store is well-nigh inexhaustible, if only it could be tapped. There is sufficient in the Sun to maintain its output of heat for fifteen billion years . . .

Aston has further shown conclusively that the mass of the helium atom is even less than the masses of the four hydrogen atoms which enter into it — and in this, at any rate, the chemists agree with him. There is a loss of mass in the synthesis amounting to 1 part in 120, the atomic weight of hydrogen being 1.008 and that of helium just 4. I will not dwell on his beautiful proof of this, as you will no doubt be able to hear it from himself. Now mass cannot be annihilated, and the deficit can only represent the mass of the electrical energy set free in the transmutation. We can therefore at once calculate the quantity of energy liberated when helium is made out of hydrogen. If five per cent of a star's mass consists initially of hydrogen atoms, which are gradually being combined to form more complex elements, the total heat liberated will more than suffice for our demands, and we need look no further for the source of a star's energy.

As this passage shows, Eddington was also a brilliant communicator, which helped him to get his ideas across — he wrote both influential textbooks and popularisations of science. This particular statement seems remarkably prescient today, since we now know that stars like the Sun do indeed generate heat by the conversion of hydrogen into helium. But that mention of just 5 percent of a star's mass consisting of hydrogen provides a clue to just how long a road astrophysicists would have to follow before reaching that conclusion—in 1920, it was still thought that the composition of the Sun was more or less the same as the composition of the Earth, and Eddington was pointing out how little hydrogen would be needed to provide the nuclear fuel required to keep it hot. It was only at the end of the 1920s that spectroscopic studies of sunlight showed that there are actually at least a million times as many hydrogen atoms in the atmosphere of the Sun as there are atoms of everything else put together (except helium), and not until the end of the 1940s that it became clear that the bulk of a star like the Sun is actually made up of about 70 percent hydrogen, 28 percent helium, and just a tiny trace of everything else. But the power of Eddington's insight, which led to the birth of astrophysics in the 1920s, was that you don't need to know where a star gets its energy from in order to describe what is going on inside it. The laws of physics tell us that a ball of gas containing a certain amount of matter and held up by the pressure inside it must have a certain size, and radiate a certain amount of energy, regardless of just where that energy comes from.

A star like the Sun does indeed behave, in many ways, like a ball of gas, obeying the same laws that apply to the air you are breathing. This is true even though the density at the heart of the Sun is many times the density of lead (but the *average* density of the Sun is only one and a half times the density of water). And the reason why matter at such extreme density behaves like a gas is that the temperature is so high that the electrons have been stripped off the atoms, leaving bare nuclei behind. In the air that you breathe, individual atoms (or molecules) fly about through empty space, bouncing off one another as they do so. Atomic nuclei are so much smaller than atoms that inside the Sun they can fly about freely, bouncing off one another and whizzing through the space between nuclei, even at such high densities (and the stripped-off electrons whiz about between the nuclei, together forming a so-called plasma). The size of a nucleus, compared with the size of an atom, is, remember, like a grain of sand in a concert hall; you can put a lot of grains of sand into a concert hall without them touching one another.

Eddington realised that there are only three things that can happen to a ball of gas in space. As it collapses under its own weight, such a ball of gas will get hot in the middle, because the gravitational energy released by the collapse makes the atoms and molecules in the gas move faster — just as Kelvin and Helmholtz explained. If the ball of gas is relatively small, it doesn't get very hot, the electrons are not stripped off the atoms, and the ball of gas settles down into a stable state, held up by the pressure of the atoms bouncing off one another, but not radiating any energy out into space — something like the planet Jupiter. If the ball of gas is very large, the heat generated by its collapse is so enormous that it makes the centre of the object so hot that the released radiation blasts the outer layers of gas away in a single huge explosion. But somewhere in the middle range of sizes — a fairly limited middle range — a ball of gas will get hot enough in the middle for electrons to be stripped off atoms and for the atomic nuclei to interact with one another, releasing enough energy to keep the star shining, but not so much that it is blown apart. Even without knowing exactly how stars maintain their output of energy, Eddington was able to use the basic laws which tell you how hot a ball of gas gets when it collapses under its own weight to calculate that a star cannot begin to glow unless it is about one tenth as massive as our Sun, and cannot hold itself together against the outward blast of radiation if it is much more than a hundred times the mass of our Sun.

One of the most interesting aspects of Eddington's application of

the basic laws of physics to stars was that it said that all stable stars, regardless of their size, must have roughly the same temperature in their
hearts. Updating Eddington's calculations slightly to take account of
modern refinements to them, the temperature at the heart of a star
must be about 15–20 million degrees Kelvin (the Sun has a central
temperature of about 15 million K; bigger stars are a little hotter
inside). If the star got any hotter than this, it would expand slightly,
which would ease the pressure on its core and make it cool down; if it
got any cooler, it would shrink a little under its own weight, releasing
heat by the Kelvin–Helmholtz process, and warming up again. Everything fitted beautifully — except that in the middle of the 1920s, when
Eddington was putting these ideas forward, nobody knew how to
make nuclear processes generate heat at such low temperatures.

The problem was that although sticking four hydrogen nuclei
(four protons) together to make one helium nucleus should indeed
release energy, each proton carries a positive electric charge, and like
charges repel one another. If two protons approach one another, even
head on, this electric repulsion will stop them from actually touching
one another, unless they are moving very fast indeed. How fast they
are moving depends on the temperature — and at fifteen million degrees they are not moving fast enough for a genuine collision to occur,
allowing the nuclear processes that make nuclei fuse (whatever they
are) to do their work.

The resolution to the puzzle came when quantum physics was
developed in the second half of the 1920s. One of the key features of
quantum physics is that on the subatomic scale entities like protons
and electrons should not be thought of as point-like particles, the way
they were thought of before about 1926, but as some combination of
wave and particle, with a fuzzy, spread-out nature. On this picture, if
a proton approaches another proton (or a positively charged nu-

cleus), the edge of the wave of the first proton can overlap with the edge of the wave of the other proton (or the nucleus) before the cores of the wave packets, as they are called, are on top of each other. The extent of this overlapping of the waves at the edges can be calculated very precisely using the laws of quantum physics, and under some circumstances is enough to allow the nuclear interactions which pull the two entities together and blend them into a single new nucleus to take place, even at temperatures like those inside the stars. It's a bit like the way you might help a child up a hill — standing on the top of the hill, you can reach out and hold on to the child's hand, pulling them up towards you. The waviness of nuclei and particles gives them a longer range of interaction, their equivalent of arm's length.

This process, called the tunnel effect, was first worked out in 1928, by the Russian physicist George Gamow, who was, at the time, interested in nuclear interactions such as alpha decay, taking place in the laboratory, on Earth. It was quickly applied to the conditions that exist inside the Sun, where it turned out to be just powerful enough to allow enough nuclear fusion to take place to provide the observed energy output from the Sun, at a central temperature in line with Eddington's calculations.

The first steps were taken in 1929, by Robert Atkinson and Fritz Houtermans. They were still thinking in terms of adding protons to larger nuclei, not the simple fusion of hydrogen nuclei to make helium, because at that time astronomers still had not realised that the Sun is mostly made of hydrogen. But they showed that at the temperatures appropriate for the heart of the Sun enough protons would indeed be moving fast enough for the tunnel effect to work some of the time. In many collisions, the proton would be repelled by the positive charge of its "target"; but the fastest moving protons could penetrate the electric barrier, as if they had tunnelled through it.

As this example shows, the developing understanding of astrophysics went hand in hand with the developing understanding of quantum physics and particle physics — it would be impossible to explain what goes on inside the Sun and stars without quantum physics, and without testing these ideas in experiments using particle accelerators here on Earth. To give you some idea of the difficulties the astrophysicists faced at first, remember that the neutron was not discovered until 1932 — and neutrons play a key role in the nuclear interactions which keep the Sun shining. It was not until the 1950s that astrophysicists were at last able to describe this process in detail, partly because they needed more information from the particle physicists, partly because it took that long to realise that the Sun is made up of more than 95 percent hydrogen and helium, and partly because of the disruption to scientific research caused by World War Two. But the story that emerged then has stood up to every further test that has been applied in the past half-century.

It is called the proton–proton (or p–p) chain, and it begins with a collision between two protons in which the tunnel effect allows them to fuse together to make a nucleus of deuterium (a deuteron), which consists of a proton and a neutron bound together by nuclear forces. In the process, they spit out a positron (which is essentially a positively charged electron, and carries away the "spare" positive charge) and a particle called a neutrino. Another proton can then tunnel into the deuteron, producing a nucleus of helium-3. Finally, when two nuclei of helium-3 interact, they can form a stable nucleus of helium-4 (made up of two protons and two neutrons bound together), spitting out two spare protons as they do so. The net effect is that four protons (four nuclei of hydrogen) have been converted into one helium nucleus, just as Eddington had suggested.

The subtle accuracy with which the modern understanding of

quantum processes explains what is going on at the heart of the Sun is brought home by looking at these reactions in a little more detail. In any gas, or a plasma like the material at the heart of the Sun, individual particles move at different speeds, spread around some average speed. The average speed goes up when the temperature goes up, but there are always some particles moving faster than the average, and some slower. For a large number of particles at a certain temperature, it is possible to calculate quite accurately what percentage of the particles will be moving at any particular speed above or below the average. Even at a temperature of fifteen million K, under the conditions that exist at the heart of the Sun the tunnel effect only allows two protons to interact in the required way if one of them is travelling at least five times faster than the average speed. Even then, the collision has to be almost head-on for the trick to work — even a fast-moving proton will not stick to another proton if it only strikes it a glancing blow.

Inside the Sun, just one proton in every hundred million is travelling fast enough to do the trick. The quantum calculations show that on average it would take an individual proton fourteen billion years to find a partner able to join it in forming a deuteron through a head-on collision. Some will take longer than average, some will find partners more quickly. The Sun is only about 4.5 billion years old, which is why most of its protons have yet to find partners in this way (in any case, only protons in the core of the Sun have any hope of taking part in the p–p chain; in the cooler outer layers of the Sun, nuclear fusion cannot occur at all). Overall, just one collision in every ten billion trillion (1 in 10^{22}) initiates the p–p chain. But there are so many protons inside the Sun, and so many collisions, that even at this incredibly slow rate, and even though only 0.7 percent of the mass of each set of four protons is released as energy whenever a nucleus of helium-4 is formed, about 5 million tonnes of mass are converted into pure energy every second at the heart of the Sun. In round numbers (to the nearest hundred million tonnes), 600 million tonnes of hydrogen are converted into 595 million tonnes of helium every second in the heart of the Sun, with the other 5 million tonnes or so being converted into pure energy. And even at this rate, so far the Sun has processed only about 4 percent of its original stock of hydrogen (again, neatly matching Eddington's calculation) into helium, even though the p–p chain has been operating for 4.5 billion years.

I have emphasised these details in order to give you a clear idea of just how well astrophysicists do understand what goes on inside stars. You cannot change these numbers by even 5 or 10 percent, and still get everything to match up. And the things that are being matched up spread across the entire range of physics, from the gas laws and the law of gravity, which describe a large ball of glowing gas, to the rules of quantum physics, which describe how subatomic particles interact with one another. The importance of all this is, of course, that the stars are the oldest things in the Universe, and the Universe itself must surely be older than the stars it contains. In order to set some lower limit on the age of the Universe, we need to know not just the age of the Earth, or the age of the Sun, but the ages of the oldest stars. And we need to be sure that our estimates are based on good physics. I hope I have convinced you that they are. The same physics that describes what goes on inside the Sun itself enables astrophysicists to estimate the ages of the oldest things in the Universe. But it isn't easy.

One of the key features in the development of the modern understanding of the lives of stars — the oldest things in the Universe — is the feedback between theory and observations. In their computer simulations (usually called models), for example, astrophysicists are able to use the understanding of how nuclear reactions take place that has been derived from particle accelerator experiments on Earth. With this calibration of their calculations, they can work out how rapidly a star like the Sun burns its fuel, and how long it will take for such a star, starting out with an initial mixture of hydrogen and helium, with just a smattering of everything else, to reach a state in which it looks like the Sun does today (there are good reasons for starting out with almost all hydrogen and helium, which we shall come on to shortly). But we already know, from the radioactive decay evidence described in Chapter 1, that the Solar System is about 4.5 billion years old. So this is another constraint on the astrophysical models — they should tell us that the model version of the Sun has a computed age of 4.5 billion years when it looks the way the Sun does today. Where necessary, the models are adjusted to make sure that this is so — perhaps by adjusting the precise initial proportions of hydrogen and helium, or the initial combination of the traces of other elements present in the model. This is not cheating — it is a way of fine-tuning the calculations to make sure that the models match reality.

Once this is done, it means that the models can be run on further into the future, to see what will happen to the Sun as it gets older still — and they can be applied to stars with different masses, to calculate how their appearance changes as time passes. Once again, the predictions of the models are checked, wherever possible, by comparing the predictions with the observed appearance of stars which are known to have different masses from that of the Sun. To take the most basic example of this kind of feedback at work, Eddington's original calculation of the physics of globes of self-gravitating gas said that no star with more than a couple of hundred times the mass of the Sun could be stable. Sure enough, no star with more than a couple of hundred times the mass of the Sun is seen in the sky, and the few that are seen with masses close to this limit are also seen to be violently active, throwing off huge clouds of material into space. This gives us confidence in the laws of physics that Eddington used to describe the behaviour of globes of self-gravitating gas, and the way in which he applied those laws. But the process didn't stop with Eddington. There has been a constant three-way interaction down the decades, as better telescopes have provided improved observations of stars, better particle accelerators have told us more about the way nuclei interact, and more powerful computers have enabled the theorists to put more and more detail into their models.

I don't have space here to go into the details of how astrophysicists unravelled the secrets of stellar evolution, which is a saga in itself (even measuring the masses of stars other than the Sun is no easy task, and involves painstaking measurements of the way stars in binary systems orbit around one another). But the story that had emerged by the end of the 1960s (and has been refined, but not altered drastically, by subsequent observations and the availability of faster computers on which to do the modelling) has one key feature which tells us, at once, that the Sun cannot be the oldest star in the Universe. The Solar System contains heavy elements which cannot possibly have been made in a star like the Sun, but must have been made in stars that were around before the Sun was born. The Sun and Solar System were made out of the debris of at least one generation of preceding stars, which ran through their life cycles relatively quickly and exploded, scattering the raw materials from which we are made out into space.

It was those first stars that were made solely out of hydrogen and helium, which, we now know, was produced out of pure energy in the Big Bang. The story of cosmology was, of course, developing alongside the story of stellar evolution in the half-century from the mid-1920s onwards, and it forms the main theme of the rest of this book.

But we have to introduce one of the most significant cosmological discoveries now. One of the key feedbacks from the theory of the Big Bang to astrophysics is the prediction, based on a combination of model calculations and the understanding of particle and quantum physics developed in experiments here on Earth, that the mixture of atomic material that emerged from the Big Bang was 75 percent hydrogen and 25 percent helium. Sure enough, the oldest stars we can see (small stars which burn their fuel slowly and have been around almost since time began) do indeed have a mixture of 75 percent hydrogen and 25 percent helium in their atmospheres, as determined by spectroscopy. They presumably have relatively more helium in their cores, where hydrogen nuclei have been fusing to make helium for billions of years; but their atmospheres are thought to contain the primordial stuff of the Universe. So we have a good idea what the first stars were made of. We also have a good idea how they made the heavy elements.

Leaving out the details, the story of the evolution of a star like the Sun can be told quite simply. The bigger (more massive) a star is, the more quickly it has to burn its fuel, because it has to generate more pressure to hold itself up against its own weight. The Sun itself has enough hydrogen in its core to maintain itself in more or less its present state for a total of about ten billion years, so it is now roughly halfway through its lifetime in its present form. A less massive star will keep on steadily burning hydrogen for longer, even though it has less to start with, because it does not need to burn so fiercely; a more massive star will have a shorter lifetime (perhaps much shorter), even though it has more fuel to start with, because it has to burn its fuel more fiercely. As you would expect, this translates into the brightness of the star. More massive stars are brighter, and less massive stars are dimmer.

The brightness of a star is also related to its colour, in the same way that a white-hot piece of iron is hotter than a red-hot piece of iron. So in a diagram (a kind of graph) where the brightness of each star (its absolute brightness, after allowing for how far away it is) is plotted against its colour, all hydrogen-burning stars lie along a single band in the diagram, a band which is called the main sequence, running roughly diagonally from top left to bottom right. The diagram itself is called the Hertzsprung–Russell, or H–R, diagram, after the two astronomers who each independently hit on this way of representing the properties of stars.

The position of a star on the main sequence depends on its mass — big, hot stars are in the top left of the diagram, and small, cool stars are in the bottom right. A star with three times the Sun's mass will stay on the main sequence for just 500 million years, and one with twenty times the Sun's mass will stay on the main sequence for only one million years. But a star with half the Sun's mass will stay on the main sequence for twenty times longer than the Sun.

Wherever a star sits on the main sequence, though, eventually all the hydrogen in its core is converted into helium. No more nuclear energy is released, so there is nothing to stop the star shrinking inwards. But when that happens, gravitational energy is released, in line with the Kelvin–Helmholtz calculation. If the star is massive enough (as the Sun itself is), the core of the star gets even hotter, to the point where a new set of nuclear reactions switch on, converting helium nuclei into nuclei of carbon. This happens at a slightly higher temperature than hydrogen burning, and the extra heat released during this phase of the star's life makes its outer layers swell up, and it becomes what is known as a red giant (red giants belong in the upper right of the H–R diagram, above the main sequence). But this is a short-lived phase of stellar evolution, which usually lasts for only 5 or

10 percent of the time the same star spent on the main sequence. When its helium fuel is exhausted, a star like the Sun will simply settle down into a more compact state (getting hot as it does so, but not hot enough to trigger more nuclear reactions) and then cool slowly, as a compact ball of stuff about as big as the Earth, but containing only a little less matter than there is in the Sun today. It will have become a white dwarf — in the case of the Sun, a ball of cooling carbon (almost literally a cinder). White dwarfs belong in the lower left of the H–R diagram, below the main sequence.

Stars that are a bit more massive than the Sun can generate enough heat in the successive stages of gravitational collapse to run through other stages of nuclear burning, in which elements such as oxygen and neon are manufactured out of carbon and helium. And if the star has more than about eight times as much mass as our Sun, at the end of its life as a red giant it collapses so violently, and so much gravitational energy is released, that two things happen. First, the gravitational energy drives a series of nuclear interactions that build up the heaviest elements — things like uranium, gold, and lead. Second, the star explodes, scattering these elements, and the ones made earlier in the star's lifetime, out into space, where they mix with clouds of hydrogen and helium gas and form the stuff from which later generations of stars (and planets, and people) are made. The star becomes a supernova, briefly shining as brightly as a hundred billion main sequence stars put together.

It is important to appreciate that all these stages of stellar evolution — main sequence, red giant, white dwarf, supernova — have been studied observationally, and the observations compared with the computer models and what we have learned from the particle physics experiments. The modern understanding of the evolution of a star is completely reliable, at the level relevant to our present story, although, of course, there are still details to be worked out and fine tuning to be done on the models. What matters, for the present discussion, is that we know that long-lived isotopes of radioactive elements such as uranium were manufactured in supernova explosions before the Solar System formed, and so these radioactive isotopes were present on Earth and elsewhere in the Solar System from the beginning of the Sun's life. And this gives us another limit on the age of the Universe.

To put this in context, the Sun and Solar System are part of a disc-shaped system of stars called the Milky Way Galaxy. It contains a couple of hundred billion stars, each more or less like the Sun, plus the clouds of gas and dust from which stars are made. We have already worked out the age of the Sun and Solar System; now, we want to know the age of the Galaxy.

One of the great things about the radioactive dating technique is that the proportions of certain isotopes around today can tell you about the proportions of radioactive elements around when the Solar System was born (now that we know the age of the Solar System), even when those original radioactive elements have all decayed — because, of course, some of the isotopes we find today, on Earth and in samples of meteoritic debris, can only have been made by the radioactive decay of specific elements that were present at the earlier time. In addition, some radioactive isotopes (such as those of uranium) are so long-lived that some of the original radioactive stuff is still around on Earth today, alongside the material that has been produced by radioactive decay of the other nuclei of the same original isotopes. This isn't quite the end of the story, because you also need to know when the radioactive stuff that went into the mixture of material from which the Solar System formed was made. But a couple of reasonable guesses can be used to give a rough idea of how old the material making up the stars in our part of the Galaxy was when the Solar System formed.

The first guess is that all of the radioactive material formed in one go, at the time the Milky Way itself formed. This is obviously wrong, since we still see supernova explosions happening today. But it is wrong in a very precisely defined and useful way. Making that assumption gives the lowest possible age for the Galaxy as a whole, and it comes out as about eight billion years. We know this number is too low; but it is important to any understanding of the age of the Universe, because it is an absolutely definite lower limit — the Galaxy *must* be older than this, and so the Universe *must* be older than this. Immediately, we can see that the Universe is at least twice as old as the Sun.

The simplest alternative guess we can make about the rate at which radioactive elements have been manufactured inside stars and distributed through the Milky Way Galaxy during its lifetime is that the rate has been steady, with the same number of supernovae going off in each millennium since the Milky Way formed. This guess probably (but not certainly) errs on the other side of the line — it seems possible that there were more supernovae when the Galaxy was young. But if you take this as an indication of a likely upper limit on the age of the Galaxy, and plug the assumption into the calculation of the relevant radioactive ages, you get a value of just under thirteen billion years, with a rather large uncertainty of plus or minus three billion years. In other words, the best that the radioactive dating technique can tell us is that the Galaxy is between ten and sixteen billion years old. The uncertainties arise both from the acknowledged limitations of the theoretical side of the technique and from the difficulty of measuring the numbers of atoms involved precisely; these sorts of uncertainties will affect every estimate of the age of the Universe that we discuss in the rest of this book. Honest scientists always quote the results of their calculations with an estimate of the uncertainties in this way, and we will only be discussing the work of honest scientists.

There is just one other number worth mentioning, before we say goodbye to the radioactive dating technique. Some supremely skilful and accurate spectroscopic observations of a single star, known as CS 22892–052, reported in 1996, have provided a measure of the proportions of the elements thorium and europium in the atmosphere of that star. This gives a radioactive age for the star itself, and the number that comes out of the calculation is 15.2 billion years, with an uncertainty of about plus or minus four billion years. To be sure, the range of ages permitted by this uncertainty is large — anything from eleven billion years to nineteen billion years. But the number is smack in the range obtained from measurements of radioactive isotopes in the Solar System, providing a crucial confirmation that the technique works. This is good news (certainly, if the two numbers had been so far apart that they didn't even overlap when allowance was made for the possible errors, that would have been very bad news, casting doubt on the whole radioactive dating technique).

That is as far as we can go using radioactive dating. But this isn't the end of the search for an age limit for the Galaxy. We still have two other techniques up our sleeve, and happily they also give numbers in the same sort of ballpark as the radioactive ages.

The first technique gives us an independent estimate of the age of the disc of material which forms the bulk of the visible Milky Way, in which the Solar System itself orbits around the centre of the Galaxy. And it brings white dwarf stars back into the story. Because white dwarf stars have no internal source of energy, and because they are solid lumps of material, and are not shrinking (so they cannot draw on gravitational potential energy to keep them warm), all they do is sit quietly in space, staying much the same size and cooling down as time passes. The rate at which they cool down can be calculated in much the same way that Buffon calculated the rate at which a ball of red-hot iron the size of the Earth would cool down, although today the physics involved is more thoroughly understood than it was in Buffon's day. To a first approximation, the rate at which such a hot object radiates energy is proportional to the fourth power of its temperature, and although there are factors to do with the internal structure of the star which alter this slightly, the theory of white dwarf cooling is well understood in terms of the basic physics.

You might think that the situation would be complicated by having many different kinds of white dwarf, with different masses from one another, to worry about. But there are sound physical reasons why no white dwarf star can have a mass bigger than about 1.2 times the mass of the Sun today—if it did, its own weight would overwhelm the pressure forces holding it up against the pull of gravity, and it would collapse still further, either forming a neutron star (a ball of neutrons, with the same density as an atomic nucleus and containing more mass than our Sun, but occupying a volume about 10 kilometres across) or a black hole. Sure enough, we do not see any white dwarf stars around with more than 1.2 times the mass of the Sun. Indeed, most white dwarfs have only about half this mass.

If a star had started its life with so little mass, it would sit on the main sequence burning hydrogen into helium for many tens of billions of years; but there is no evidence that the white dwarfs we see today always had the mass they have now (for example, their spectra reveal the presence of elements heavier than helium, which can only have been made inside a giant star). The astrophysical models tell us that stars up to about eight times the mass of the Sun eject matter during the later stages of their lives, when they are giant stars, and end up with less than the critical mass required to make a stable white dwarf.

Again, theory is confirmed by observation, which shows just such massive stars blowing great clouds of material away into space. The white dwarfs we see today are the cores of massive red giants which have lost their outer layers entirely. Stars with even more than eight times the mass of the Sun cannot eject enough mass to end up as white dwarfs, but that is another story. What matters here is that stars at the top end of the range which can make white dwarfs, with nearly eight solar masses of material in them, run through their main sequence lifetimes very quickly, in the span of only a few million years. So the first white dwarfs formed when the Milky Way Galaxy was very young. Even better, since they all formed as the cores of red giants, they all formed at the same temperature — a temperature we know accurately from the comparison between models, observations, and particle experiments. And since they are all roughly the same mass, they all cool at roughly the same rate. If we can find them today and take their temperatures, we will have another lower limit on the age of the Milky Way.

The snag is that by definition the oldest white dwarfs will be the faintest, since they have cooled most. So they will be the hardest to find, because they will be dim. We can never be absolutely certain that we have found the faintest white dwarfs there are, which is one reason why this is only a lower limit on the age of the Galaxy (the other is that it must have taken some time, even if only a few million years, for the progenitors of these stars to evolve to the white dwarf state). The actual cooling ages of the faintest white dwarf stars yet detected are about 9.5 billion years, which, adding in a reasonable allowance for the time it took for them to form in the first place, sets a lower limit on the age of the disc of the Milky Way of about ten billion years. This is exactly right to match the estimates of the overall age of the material in the disc of the Galaxy inferred from the radiometric technique.

This is especially pleasing because it uses a completely different technique. It was good news that the radiometric age of material in the Solar System is the same (within the error limits) as the radioactive age of CS 22892–052, because the same technique applied to two different objects in two different places gave the same answer, which suggests that the technique works. Now, we have applied a completely different kind of technique to a completely different kind of object, and still got an answer consistent with the radiometric age of the disc. You don't need to know anything about radioactivity to measure these cooling ages, and you don't need to know anything about cooling to measure the radioactive ages. Given that the radiometric ages check in above ten billion years but below fifteen billion years, if we had found that the ages of the oldest white dwarfs were only one billion years, or as great as 100 billion years, we would have been deeply worried; the fact that we get ten billion years out of the calculation suggests, however, that we are doing something right, with both techniques. And there is still a third way to estimate the age of the Galaxy.

Remember that in the H–R diagram, more massive stars lie at the top left of the band corresponding to the main sequence. But more massive stars are also stars which have shorter lifetimes on the main sequence. If we had a group of stars which were all the same age, and we watched them for tens of millions of years, plotting out their positions in the H–R diagram from time to time, we would find that although we started out with a fine broad main sequence stretching grandly across the diagram, after a few million years the stars in the top left of the diagram would be disappearing, as they used up their hydrogen fuel and left the main sequence — but not disappearing entirely, as the same stars would now show up further to the right in the diagram, as red giants. At any given time, the point where the main sequence now ended, and bent to the right to link up with the region of the diagram known as the red giant branch, would depend only on the masses of the stars at that bend in the sequence; and the masses of the stars at that point would tell you how old they were, because the only thing that determines how long a star stays on the main sequence is its mass. Because we know from comparison between models and observations of many stars what mass corresponds to a particular position on the main sequence, all we would have to do would be to measure the position of the bend in the main sequence to know how old the stars were — *provided,* that is, we knew they were all the same age.

If you look out into the Milky Way at large, this is a useless way to measure ages, because the stars you see were all born at different times. But there are places, outside the plane of the disc of the Milky Way, where large numbers of stars — millions of stars, in some cases — all formed together, out of one cloud of collapsing gas. They are called globular clusters, because they form spherical concentrations of stars, closely packed together. In the denser regions of such a globular cluster, there may be as many as a thousand stars in each single cubic parsec of space (a parsec is just over 3.25 light-years); for comparison, there is no other star within one parsec of our Sun. On average, the spacing between stars in a globular cluster is about one tenth of the spacing between stars in our part of the Milky Way. There is no doubt that all the stars in an individual globular cluster did indeed form at the same time (or within a few million years of each other) from a single cloud of gas. Altogether, about 150 of these star systems are known, and they are spread in a spherical halo around the disc of the Milky Way; but most of them are no further from the centre of the Milky Way than we are (a distance of about 9 kiloparsecs, or 30,000 light-years). Spectroscopic studies of the stars in these globular clusters show that their atmospheres are almost entirely composed of hydrogen and helium, the raw material which emerged from the Big Bang, and it is clear that they formed in the first stage of the collapse of the huge cloud of stuff that eventually settled down to form the Milky Way itself. So the globular clusters are expected to be older than the disc of the Milky Way (which includes the Solar System), and are probably the oldest things whose ages we might, in principle, measure directly.

If all the stars in a globular cluster formed at the same time, then over the eons the more massive stars have successively used up their hydrogen fuel and formed red giants (and then white dwarfs, neutron stars or black holes). The bend in the H–R diagram steadily creeps down the main sequence as time passes. If we had a large auditorium full of people with different heights, adults and children of all ages, and someone in charge who plotted out the number of people with each different height on a blackboard every half-hour, we could make a similar time-dependent graph. Every half-hour, the person in charge would send all of the people above a certain height out of the room, and rub out their contributions to the graph. Each half-hour, the height limit would be reduced by a predetermined amount — maybe a couple of centimetres. The line on the graph would shrink away as time passed, and anyone who knew the rules could look into the room and check the graph at any time, and know from the point where the line stopped how long the experiment had been going on (at least, to the nearest half-hour).

In principle, the same sort of reasoning can be applied to the H–R diagrams for globular clusters. But remember that the H–R diagram is based on measurements of the brightnesses of the stars involved. The apparent brightness (or apparent magnitude) of a star in a globular cluster depends on its distance from us, and if you just used the apparent brightnesses of the stars in making the H–R diagram you would get an incorrect estimate of the age of the cluster by this technique. You have to know the distance to the cluster, from some independent technique, so that you can calculate how bright the stars would be if they were being viewed from close up (this correction is always made to correspond to viewing from a distance of 10 parsecs, and the resulting corrected brightness is called the absolute magnitude). But how do you measure distances to globular clusters?

This question touches on the central theme of the rest of this book, measuring distances across the Universe in order to work out the age of the Universe. The development of the first cosmic distance scale is described fully in the next chapter; the discovery that matters here is that there is a class of stars, known as RR Lyrae stars, which each vary in a regular way, brightening and dimming in a characteristic fashion over a period of a few hours. Several thousand RR Lyrae stars have been seen in the disc of the Galaxy, and more than 1,500 are known in globular clusters. Because the distances to some RR Lyrae stars can be worked out by other techniques (more of this later), astronomers have known for some time that they all have roughly the same average intrinsic brightness (roughly the same absolute magnitude, averaging over a complete period of variation). This already gave a rough guide to the distances of globular clusters, from measurements of the apparent brightnesses of the RR Lyrae stars they contain; but more recently it has been discovered that there is a very close relationship between the exact colour of an RR Lyrae star and its absolute brightness. So if you can measure the colour of such a star in a globular cluster (and by "colour" astronomers mean a precise measurement of the brightness of the star at each of a series of precise wavelengths, across the spectrum), then from its apparent magnitude you can work out how far away the cluster is rather more accurately.

There is also a cruder technique for estimating the distances to globular clusters, which involves taking the H–R diagram for a cluster and adjusting all the magnitudes in it (in effect, sliding the line representing the cluster's main sequence around on the H–R diagram) until the main sequence lies right on top of the standard main sequence determined from studying nearby stars. The amount of adjustment you need to make ought to tell you how far away the cluster is; but the snag with this is that nearby stars are, like the Sun, made up from material which has already been processed in one or more generations of earlier stars, so their chemical composition is different from the chemical composition of the primordial stars that make up the globular cluster population. This definitely affects the position of the main sequence in the H–R diagram, but nobody is quite sure just how it affects it. So the RR Lyrae technique is the best one for measuring the distances to globular clusters, and thereby (with the aid of the bend in the main sequence) inferring their ages.

Don't, though, run away with the idea that this is easy; the description of the whole technique that I have outlined here glosses over a wealth of observational difficulties. Indeed, you can get some idea of just how difficult a task all this is from the fact that even by the late 1980s, the distances to only about ten globular clusters had been determined with any degree of accuracy. After all, remember that the observers are trying to determine the precise amounts of energy being radiated at precise wavelengths in the spectrum from individual stars in clusters that may contain hundreds of thousands of stars, at distances typically of thousands of light-years. As we shall see later, things have since improved dramatically, not least through the observations made by the Hipparcos satellite, but also thanks to the development of new electronic detector devices (CCDs) for use with ground-based telescopes. But I don't want to go into all these new developments until I have brought you up to date with the story of our understanding of the Universe at large.

There has been a lot of debate about just how to interpret the observations of globular clusters to give us ages, and there is theoretical uncertainty, as well as observational uncertainty, because the models are not yet good enough to mimic the evolution of these systems as accurately as we would like. The bottom line, though, is that the pre-Hipparcos estimates of the ages of the oldest globular clusters associated with our Galaxy come out at around fifteen billion years, with an uncertainty of about plus or minus three billion years. If the pre-Hipparcos estimates of the distances to globular clusters are correct, it would be extremely difficult to push the lower end of the range of these estimates down beyond twelve billion years. Which seems fine — after all, our estimates of the age of the disc of the Milky Way all point to a minimum age of ten billion years, and we expected the globular clusters to be older than this.

In the mid-1990s, it seemed that the oldest things in the Universe were about fifteen billion years old. Which means that the Universe itself must be at least fifteen billion years old. But how *do* you measure the age of the Universe? It all starts, as I have hinted, with measurements of cosmic distances — distances beyond the Milky Way Galaxy in which the Sun, all the stars of the Milky Way itself, and the globular clusters live.

3

Across the Universe

The First Cosmic Distance Scale

In order to understand how astronomers developed the skill to measure distances across the Universe to other galaxies (indeed, to understand how they developed the skill to recognise other galaxies for what they are), we need to take a step backwards in time and closer to home, to look first at how they measure any distances to objects beyond the Earth. At first sight, it is almost as remarkable that we can measure the distance to the planet Mars as it is that we can measure the distance to another galaxy — and measuring the distance to Mars is, as we shall see, a crucial factor in determining the distances to the galaxies.

It all starts with the most basic technique used by surveyors — triangulation. If you want to know how wide a river is, say, without getting your feet wet, one way to do this is to choose a distinctive feature (such as a tree) on the far bank and to measure out a baseline on your side of the river. From each end of the baseline, you observe the tree through a theodolite (essentially a small telescope mounted on a tripod) and measure the angle between the line of sight and the baseline. With both these angles and the length of the baseline known, it is a simple matter to calculate the distance of the tree from the centre of the baseline using the rules of geometry. In a familiar variation on the theme, the surveyor gets an assistant to stand at the point of interest (it doesn't matter if the assistant gets wet feet) holding a rod of a known length. In this case, the theodolite is kept in one place, but is focused first on one end of the rod, then on the other end, and the angle between the two lines of sight is noted. Once again, knowing the length of the base of the triangle (the length of the rod) and the relevant angle (in this case, the angle of the apex of the triangle), it is easy to calculate the distance from the theodolite to the centre of the baseline (the rod). And it is easy to see why the technique is called triangulation.

If you have a long enough baseline, you can measure the distance to anything. The Moon is our nearest neighbour in space, and it is fairly straightforward to make observations of the Moon at the same time from widely separated observatories — perhaps on either side of the Atlantic Ocean. You don't even need the telephone to make sure the observations are carried out simultaneously. Each observatory notes the position of the Moon against the background of stars at some prearranged time, and this tells the observers at each observatory the angle between the line of sight to the Moon and an imaginary line joining the two observatories. Triangulation in this way reveals that the distance to the Moon is about sixty times the radius of the Earth (you then have to measure the radius of the Earth, of course, and that involves more surveying techniques which we won't go into here; the bottom line is that the distance to the Moon, averaged over a month, is 384,400 kilometres). So the triangle used in triangulating the distance to the Moon is about sixty times taller than the baseline used in the triangulation is wide — a pretty tall, thin triangle.

The fact that astronomers can measure the distance to the Moon in this way today may not seem very impressive. But it is rather more impressive that as long ago as 1671 astronomers used exactly the same technique to measure the distance to another planet, Mars. It involved one team of observers making observations of the position of Mars against the background stars from the comfort of the Paris Observatory, while another team travelled to Cayenne, in French Guiana, to make simultaneous observations of the red planet. The great thing about measuring the distance to Mars (or, indeed, to any other planet in the Solar System) is that this is the only extra piece of information you need, along with the known orbital periods of the planets (the time it takes for them to orbit around the Sun once), to be able to calculate the distance from each of the planets to the Sun, using the laws of planetary motion discovered by Johann Kepler at the beginning of the seventeenth century and explained by Isaac Newton's inverse square law of gravity. Indeed, you don't actually need to know the explanation for Kepler's laws (Newton didn't publish his theory of gravity until 1687) in order to use them in this way. The French astronomers determined that the distance from the Earth to the Sun is 140 million kilometres — only about 10 million kilometres less than the best modern estimates.

Those modern estimates no longer depend on triangulation to work out the distances to the planets, and thereby the distances to the Sun. We can bounce radar echoes off Venus and Mars, the two planets with orbits nearest to the orbit of the Earth, and measure their distances from the time it takes for the radar pulses, travelling at the speed of light, to get there and back again. But triangulation is still crucial to the modern understanding of the cosmic distance scale, because all of this messing about with the geometry of the Solar System gives us one key distance, the average distance from the Earth to the Sun, with very great precision. It is 149,597,870 km, and the possible error in the determination of this particular astronomical distance is less than ±2 kilometres. Twice that distance, the diameter of the orbit of the Earth around the Sun, is the longest baseline that any human observer has yet been able to use for triangulation — although some astronomers dream of putting an automatic observatory in orbit around the Sun beyond the orbit of Jupiter, with a correspondingly greater baseline to work with. So far, though, the diameter of the Earth's orbit is the best baseline that we have, and the one that gives us the only direct measurements of the distances to the stars.

Astronomers were very quick to realise the potential of these kinds of measurements for determining distances across the Solar System. Remember that the idea of a Sun-centred Solar System was only properly worked out by Nicolaus Copernicus in the early decades of the sixteenth century, and not published until the year he died, 1543. The idea only gained widespread acceptance after the theoretical work by Kepler (who discovered that the orbits of the planets around the Sun are elliptical, not circular), and the observations of Galileo, who found powerful evidence that the Sun-centred model was right — one of the most impressive of his discoveries was that Jupiter is orbited by four large moons, in much the same way that the planets orbit the Sun.

Kepler's laws of planetary motion (which also apply to the orbits of moons) were only worked out in the first decade of the seventeenth century, and Galileo's discoveries were described in a book, *The Starry Messenger,* published in 1610. The idea that the Earth orbited the Sun was still, literally, heretical at that time (at least, in the eyes of the

Catholic church, which convicted Galileo of heresy in 1632 and confined him to house arrest for the rest of his life). And yet, as we have seen, by 1671 astronomers were using this "heretical" knowledge to calculate the width of the Earth's orbit around the Sun.

Galileo's observations with the first astronomical telescopes also sounded the death-knell of the idea that the stars might be tiny lights attached to a crystal sphere, not much further from the Sun than Saturn (the most distant planet known to the ancients). But just how far away were the stars?

The gap (both scientifically and historically) between Copernicus and Kepler was filled by Tycho Brahe, who lived from 1546 to 1601, and made superbly accurate observations of the changing positions of the planets on the sky, which were used by Kepler in working out his laws of planetary motion. But, like most people at that time, he did not accept the Copernican idea of a Sun-centred Solar System. One of the reasons why he did not accept it was because the stars could not be seen to move over the course of a year. If the Earth were really in orbit around the Sun, he reasoned, then at intervals of six months apart we would be looking at the stars from either end of a baseline equal to the diameter of the Earth's orbit. Any surveyor could tell you (and there were good surveyors around in the sixteenth century) that this would change the angle at which you viewed distant objects, so that they would appear to shift across the sky. The effect is called parallax, and you can see it at work just by holding up a finger at arm's length, and closing each of your eyes alternately. The finger seems to jump across the background view, because you are viewing it from either end of a baseline equal to the distance between your two eyes. The stars did not show a parallax effect, so Tycho, and others, reasoned that the Earth could not be moving. Or at least, if the Earth were moving, then, Tycho calculated, the stars must be at least seven hundred times further away than the most distant planet. This seemed inconceivable to him, so he rejected the idea.

After the work of Kepler and Galileo, though, the puzzle resurfaced. As people became increasingly convinced that the Earth did orbit around the Sun, and especially after the diameter of the Earth's orbit was measured in 1671, the failure of the stars to show any parallax became increasingly uncomfortable. Could the stars *really* be so far away that they showed no parallax even using a baseline 300 million kilometres (in modern units) long? There was, though, another clue that the stars must be at immense distances — their faintness. As people gradually came to consider the idea that the stars might be objects like the Sun, but at great distances, so they began to try to work out just how far away an object like the Sun would have to be in order to be as faint as a star. It is a useful convention in astronomy to measure distances in terms of the average distance from the Earth to the Sun, which is called the astronomical unit, or AU. This means that you don't have to worry unduly about whether the distance to the Sun is 140 million kilometres, or 150 million kilometres, because you can always plug the exact numbers in at the end of the calculation. To give you some idea of the scale of the Universe as it was understood before the invention of the astronomical telescope by Galileo, the distance from the Sun to Saturn is just 10 AU.

One of the other great advances that was made in the seventeenth century, alongside the discoveries about the nature of the Solar System, was in the study of optics. Both Isaac Newton and the Dutch physicist Christiaan Huygens investigated the nature of light, and one of the things they knew was that the intensity of a light source falls off as the square of its distance — double the distance, and the light will seem only one quarter as bright, and so on. Huygens, who lived from 1629 to 1695, hit on a way to use this to measure the distance to the bright star Sirius. If you knew, for example, that Sirius seemed to be only one millionth as bright as the Sun, and if you guessed that Sirius was really just the same brightness as the Sun, then you would infer that Sirius is a thousand times further away than the Sun is (you would place Sirius at a distance of 1,000 AU in your map of the cosmos), because a thousand is the square root of a million.

Huygens tried to compare the brightness of the Sun and Sirius by allowing sunlight to enter a room through a tiny hole in a screen, and making the hole just the right size to make the light that got through it as bright as Sirius. If he could measure the fraction of the Sun's surface that was visible through the hole, he could work out what fraction of the Sun's brightness corresponded to the brightness of Sirius (which is the brightest star in the night sky).

This was difficult enough in itself, but remember that there were no photographs in those days, and the comparison of the two bright-nesses had to be made from memory. In all honesty, the method was very rough-and-ready. But it gave Huygens an estimate of 27,664 AU for the distance to Sirius, one of the first pieces of direct evidence that the Solar System is only a tiny, insignificant speck in a much greater cosmos.

A slightly more cunning, and more accurate, technique was de-vised by the Scottish mathematician James Gregory, and published in 1668. Gregory pointed out that it would be easier to compare the brightness of Sirius with the brightness of one of the planets — indeed, you could choose a time in the orbit of a planet when its brightness, as seen from Earth, closely matched the brightness of Sirius. A planet is visible only because it reflects light from the Sun, and Gregory could calculate how bright the Sun would seem from the distance of the planet, estimate how much of the arriving light would be reflected, and calculate how bright this reflected light would look by the time it got back to Earth. Putting everything together, he came up with a distance of 83,190 AU for Sirius.

But this estimate had to be revised upwards when the scale of the Solar System, including the distances to the planets used in the test, was recalibrated. Isaac Newton himself updated Gregory's calculation, and came up with a distance for Sirius of 1 million astronomical units — but he left the startling result to languish in the pages of the draft version of his *System of the World,* which was only published in 1728, the year after his death. Astronomers at last became aware of the true immensity of the Universe, and understood why stellar parallaxes had not been noticed, even using a baseline as long as the diameter of the Earth's orbit around the Sun (2 AU). And it was to be more than a hundred years after the publication of the *System of the World* before telescopes and observing techniques became accurate enough to measure the distances to a few nearby stars directly, using parallax.

Galileo had pointed out the best way to do this, in his *Dialogue on the Two Great World Systems,* which was published in 1632 (and was the immediate cause of his trial for heresy). If two stars happen to lie close to one another along the line of sight from Earth, but one is much further away than the other, then over the course of a year the nearer star will seem to move to and fro compared with the distant star, because of parallax — exactly as your finger seems to move to and fro against the background when you look at it with each eye alternately. And because you are comparing two stars along nearly the same line of sight, many of the problems astronomers have to cope with, such as the effect of the atmosphere of the Earth on the light passing through it, should cancel out from the comparison.

When such comparisons could at last be made, at the end of the 1830s, they showed just how far away even the nearer stars are. The actual parallaxes measured are tiny — less than 1 second of arc. To put that in perspective, there are 360 degrees in a circle, 60 minutes of arc in each degree, and 60 seconds of arc in each minute. The size of the Moon on the night sky is just over 30 arc minutes; the size of the *largest* parallax displacement observed for any star is less than one sixtieth of one thirtieth of the size of the Moon seen from Earth — about one two-thousandth of the apparent diameter of the Moon.

In order to convert these angular measurements into distances, astronomers like to keep with the astronomical unit as their yardstick, which is easy, since the AU is just half of the baseline used in their observations. They define one "parallax second of arc," or parsec, as the distance from which a baseline 1 AU long would subtend an angle of 1 arc second. So a star 1 parsec away would show an angular shift across the sky in the course of six months of 2 seconds of arc, since the diameter of the Earth's orbit is 2 AU. If you put the best modern measurements of the distance from the Earth to the Sun in, 1 parsec corresponds to a distance of 3.2616 light-years, where a light-year is the distance that light, travelling at 299,792 kilometres per second, could travel in one year (so a parsec is just over 30,000 billion kilometres). There is no other star within 1 parsec of the Sun; the nearest, Alpha Centauri, is 1.32 parsecs (4.29 light-years) away. Sirius, which was studied by Gregory, Huygens, and Newton (they picked Sirius precisely because it is the brightest star in the sky and they guessed that this meant it must be relatively close to us), is indeed one of our nearest neighbours, at a distance of 2.67 parsecs (8.7 light-years). One astronomical unit is just 499 light seconds (it takes light 499 seconds to reach us from the Sun), so the distance to Sirius is just under 550,000 AU. This is impressively close to the 1 million AU estimated by Newton, although his estimate was not really that accurate, and it just happens that some of his incorrect guesses cancel each other out (for example, Sirius is actually much brighter than the Sun, and if Newton had known this he would have placed it much further away). To give you some idea of how difficult it is to obtain parallax measurements, it took until 1878 for the distances to just seventeen stars to be measured in this way, and even by 1908 only about a hundred stellar parallaxes had been measured.

Once you know the distance to a star, you also know, from the inverse square law, how bright it really is — its absolute magnitude, defined as the brightness it would have at a distance of 10 parsecs ("magnitude" is the unit in which astronomers measure brightness, so the absolute magnitude of a star is a precise number, like its temperature). Because the apparent magnitude depends only on the absolute magnitude and the distance, astronomers sometimes describe distances to stars in terms of what they call the distance modulus, which is just the difference between the apparent magnitude of a star and its absolute magnitude.

The most important feature of stars that was discovered from this determination of their absolute magnitudes and distances was the main sequence of the H–R diagram. This discovery had to await the measurement of enough distances, and "enough" is about a hundred, so it is no coincidence that Hertzsprung and Russell each independently made the discovery early in the twentieth century. By the end of the 1970s, the distances to all the stars within 22 parsecs of the Sun (just under 500 of them, all with parallaxes greater than 0.044 seconds of arc) had been determined, and the plot of their brightnesses (absolute magnitudes) compared with their colours (determined in the usual precise way in terms of brightness at particular wavelengths) clearly shows the main sequence and a scattering of white dwarfs in the bottom left of the diagram. Before Hipparcos, the best catalogue of stellar parallaxes was one compiled by the Yale University Observatory, with information on about 7,500 stars, but with large uncertainties in many of the measurements. But how can we determine distances further out into space?

Parallax measurements can only give the distances to the nearer stars, even with modern instruments; but they did give the first clue to the distance scale of the Universe. From the middle of the nineteenth century onwards, astronomers have developed various other techniques to measure distances to stars too far away for their parallaxes to be measured. Some of these techniques seem to be fairly reliable; others are more approximate. But none of them is a direct, geometrical measurement of distance in the way that triangulation is, whether or not it goes by the name of parallax. Once we move beyond the range of parallax, there will be an element of uncertainty in every distance measurement we use.

One very straightforward technique uses the main sequence relationship. If we can measure the colour of a star, and if we think that it is a main sequence star, then we know its absolute magnitude, from its colour. So all we have to do is measure its apparent magnitude to work out its distance. There are several snags, though, even with this simple technique. First, the colour of the star may be affected by the presence of clouds of gas and dust along the line of sight; then, the overall apparent brightness of the star may be affected by this material. You can see this process at work for yourself. Dust along the line of sight tends to scatter away blue light, and let red light through, so the colour of the star behind the dust is made redder, and the image is fainter. This is why the Sun looks red and dim at sunset — because its light is being scattered by dust in the atmosphere of the Earth. Exactly the same sort of reddening and dimming (grandly dubbed interstellar extinction) occurs in the light from distant stars as it passes through clouds of material in space (this interstellar reddening, of course, is nothing to do with the redshift, which we will come to soon).

Even if we have some way of allowing for the reddening, we might have guessed wrong in the first place — the star might not really belong on the main sequence at all. More sophisticated versions of this kind of approach relate the absolute magnitude of a star to more subtle details of its visible spectrum of light. But all of these techniques have one thing in common — they are more reliable if the rules on which they are based (the main sequence, or the subtle spectroscopic signatures) can be calibrated using larger numbers of stars whose distances have been determined by other means.

There are two of those "other means" that are crucial to our understanding of the distance scale of the Universe and its age. Both depend on being able to measure the speeds with which stars are moving, and that in turn depends upon the way the spectrum of light from a star is affected by its motion — the Doppler effect.

Light travels through space in the form of a wave, like ripples on a pond. A particular colour of light corresponds to a particular wavelength of light, with the peaks and troughs in the ripples spaced a certain distance apart. If the source of the light is moving towards you, though, the ripples get squashed up in front of the moving object by its motion, which means the light has a shorter wavelength. In the visible spectrum of light, which runs from red through orange, yellow, green, blue and indigo to violet, longer wavelengths are at the red end of the spectrum and shorter wavelengths are at the violet and blue end of the spectrum. So this shift in the wavelength of light from an object moving towards you is called the blueshift (it ought to be called the violet shift; I don't know why it isn't). Similarly, if an object, such as a star, is moving away from you, the waves it leaves behind are stretched by its motion, so the features in the spectrum move to longer wavelengths; there is a redshift.

If stars just emitted a uniform spread of light at all colours, these effects would not be noticeable. In fact, though, the spectrum from a star, revealed by passing its light through a prism and splitting it into its rainbow components (the technique known as spectroscopy), contains sharply defined light and dark lines, which correspond to the presence of particular elements in the atmosphere of the star (this is how we know what stars are made of). Those sharp lines are always produced at precisely known wavelengths, identified from studies of the light from hot objects in the laboratory on Earth; so by measuring the shift in those lines in starlight towards the red or blue (violet) ends of the spectrum astronomers can tell not only whether a star is moving directly towards us or away from us, but how fast it is doing so.

This is only half the story, because the star is virtually certain to be moving across the line of sight, as well. Its true velocity through space can only be worked out by measuring the Doppler velocity (the speed with which the star is moving directly towards or away from us) and its sideways velocity (the speed with which it is moving across the sky) and adding the two components together in the appropriate way, to find out its actual speed through space and the direction it is moving in.

Just as relatively few stars are close enough for their distances to be measured directly by triangulation and parallax, so relatively few stars are close enough for their motion across the sky to be measured. But this time there is at least one thing the observers have going for them. The parallax of a star is something you measure over the course of a few months, and it will be the same from one year to the next. But the motion of a star across the line of sight (its proper motion, in astronomical jargon) adds up from year to year. Provided its position is measured accurately at the start, the longer you wait the easier it will be to tell that the star has moved against the background of more distant stars. This is powerfully highlighted by the way proper motion was discovered. Edmund Halley was interested in cataloguing the positions of the bright stars, and in 1718 he noticed that the positions of three stars recorded in old catalogues compiled by the ancient Greeks were not in the places the Greeks had seen them. Sirius, Procyon, and Arcturus had all moved noticeably across the sky in the space of a couple of millennia — Sirius by a full degree of arc, twice the diameter of the full Moon as seen from Earth.

It is no coincidence that these three stars turned out to be very close to us, nor that they are among the eight brightest in the sky. The closest stars are likely both to look bright and to be moving visibly across the sky. But the three properties do not always go hand in hand. The largest observed proper motion is for an object known as Barnard's star, which is only 1.8 parsecs away from us and races across the sky at 10.3 seconds of arc (half of 1 percent of the Moon's angular diameter) each year; but it is much too faint (with an absolute magnitude only one hundredth of that of the Sun) to be seen by the naked eye. The typical proper motion for a star visible to the unaided human eye is less than one hundredth of the record set by Barnard's star.

Using velocities to determine distances to stars depends on having a group of stars (the more the better) all at roughly the same distance from us, moving together through space. Fortunately, there are groups of stars like that. They are called clusters, and the ones that are relevant to the present part of the story are called open clusters, because they have a loose structure, quite different from the tightly packed globular clusters that we have already met. The moving cluster method of determining distances depends on measuring the transverse velocities — the proper motions — of all the stars in the cluster.

The first thing that matters is the direction across the sky in which each star seems to be moving. Because the cluster is moving together through space, when you draw a line on a star map showing the direction each star is moving all of the lines seem to be converging on a single point on the map (or diverging from a point behind them). The stars are not really converging in this way; it is an optical illusion, just like the optical illusion which seems to make the parallel lines of a long, straight railroad track meet at a point on the horizon. What matters is that this tells you the true direction in which the cluster as a whole is moving through space. And by measuring the Doppler velocities, you know the actual velocities of the stars (all of them much the same as each other because they are in a cluster) along the line of sight (their radial velocities). From these two measurements, you can work out what fraction of the actual motion of the cluster as a whole, in kilometres per second, corresponds to its proper motion across the sky, measured in seconds of arc. Now it is time to bring the speed with which the stars seem to be moving across the line of sight into the calculation. If you know the real sideways (transverse) velocity, you know how far the star has actually moved at right angles to the line of sight in a year, or a decade, or longer; this becomes your baseline. And if you know the measured angular shift of the star across the sky in that time, you can work out the distance, once again, by triangulation.

This is a very nice trick, which is also, like parallax, a direct geometrical technique that represents a real measurement of distances. But it is more unreliable than parallax, because the stars in the cluster are not all moving precisely in parallel lines towards a point on the horizon, so to speak; they are affected by each other's gravity, and by other factors, so that they each have a component of random velocity, as well as the velocity they share as members of the cluster. And, of course, by its very nature a cluster is a spread-out thing. The best you can hope for is to measure the average distance to the cluster, and individual stars in the cluster will be a little bit closer to us, or a little bit further away, than the average.

The technique works if you have a cluster that has plenty of stars in it, and if it is close enough for those individual stars to be spread out across a reasonably large patch of sky, so that the directions of their individual motions can be clearly picked out. The bad news is that there is only one cluster, called the Hyades cluster, which actually meets these requirements. The distances to a couple of other clusters have been estimated roughly using the technique. But the bottom line is that we only know the distance to a single cluster of stars reasonably accurately from direct geometrical methods. The moderately good news is that the Hyades cluster is just on the limit of being probed by pre-Hipparcos parallax methods, and although the uncertainties in these measurements are large at such distances (so they could not be depended upon on their own), they do give a distance which matches the moving cluster method distance. And the good news is that the Hyades cluster contains a lot of stars (several hundred, spread over a volume revealed by this technique to be a bit more than 10 parsecs in diameter) with different properties, so that by measuring the distance to this one cluster we can calibrate the spectroscopic techniques which relate the visible appearance of a star to its absolute magnitude. Even better, there doesn't seem to be any extinction or reddening of the light from the cluster, because there is no obscuring material in the line of sight.

The conclusion is that the Hyades cluster has a distance modulus of between 3.2 and 3.4, corresponding to a distance of 40 to 50 parsecs. One word of warning, though. Over the forty years from 1940 to 1980, the accepted "best estimate" of the distance to the Hyades cluster went up from 35 parsecs to 45 parsecs, an increase of 30 percent. And this is the easiest distance measurement that we have to worry about when working out the cosmic distance scale.

I'm almost ashamed to tell you about the other crucial technique for measuring distances to stars, because it sounds so silly. But it does work, after a fashion. The idea is that all the stars in our neighbour-hood of the Milky Way are moving together around the centre of our Galaxy, more or less in the way that a planet like Jupiter and its family of moons is orbiting around the Sun together. If this were exactly the case, then the distances to the nearby stars could be worked out by measuring their proper motions and their radial velocities, and setting them at the appropriate distances for these velocities to add up to give the stars the same overall motion around the Galaxy that we have. But of course, individual stars have random motions, because they are influenced by the gravity of nearby stars, and by the dynamics of the clouds of gas from which they formed; they are not all moving on perfectly circular parallel railroad tracks around the Galaxy. So this is a pretty hopeless way to measure the distance to an individual star. But it turns out to work quite well in giving you the average distance to a large number of stars scattered all over the sky, in which the individual random velocities cancel out (we know it works quite well, because wherever possible, distances determined in this way are compared with distances determined by other techniques).

The technique is called statistical parallax, and it can be used to give distances up to about 500 parsecs away. One way you might use this technique would be to take a sample of stars which all have the same colour but lie in different parts of the sky, and work out an "average distance" to them. This would give you an idea of the average abso-lute magnitude of a star with that colour. Then, if you see a star which is the same colour but too far away for its distance to be measured di-rectly using any of these techniques, you can guess that it has the same absolute magnitude as the average you have just worked out, and de-termine its distance by comparing this with its apparent magnitude.

This is a fairly rough-and-ready guide if you actually do use colour to select your sample of stars, because many factors can affect the perceived colour of an individual star. But statistical parallax has been particularly important historically in determining distances to RR Lyrae variables, which, as we have seen, can then be used as distance indicators in their own right. And, as I have already described, once the distance to a single open cluster of stars, the Hyades, was determined reasonably accurately, the distances to other open clusters could be worked out by comparing the H–R diagrams for those clusters with the H–R diagram for the Hyades, and adjusting the brightnesses of the stars to make the main sequences lie on top of one another (although this is where problems of reddening and extinction rear their ugly heads). Main sequence fitting, as it is called, works reliably for young massive stars (which are hot and bright) out to distances of about 7,000 parsecs (7 kiloparsecs). But globular clusters, remember, have different H–R diagrams from open clusters, because they contain a different kind of star, formed when the Galaxy was young.

Unfortunately, although RR Lyrae stars are good distance indicators, and were particularly important in working out the distances to globular clusters and thereby the scale of our own Galaxy, they are too faint to be much use in determining the distance scale out into the Universe at large. The key stars that were used to take us beyond the Milky Way are much brighter variables, called Cepheids. And to understand their importance to the story, we have to go back to the early years of the twentieth century, before anyone even knew how a star worked, and pretend that we don't know about things like the distances to globular clusters and the way the Sun orbits around the centre of the Milky Way.

The first glimmer of an understanding of what the Milky Way is had come only at the beginning of the seventeenth century, when

Galileo turned his first telescope on it. We are so used to the idea that the Sun is just one star among hundreds of billions of similar stars which together make up the Galaxy called the Milky Way that it is hard to put Galileo's discovery in perspective. But perhaps it will help if I point out that on the darkest, blackest night, with no Moon in the sky and no clouds to obscure the view, on a tall hill far from any city lights, the largest number of stars that you could pick out with the naked human eye, anywhere on Earth, would be only a couple of thousand. It is in that context that you have to place Galileo's remark that when he turned a telescope on the Milky Way he saw stars "so numerous as almost to surpass belief," and wrote in his book *The Starry Messenger* that the Milky Way: "is in fact nothing but congeries of innumerable stars grouped together in clusters. Upon whatever part of it the telescope is directed, a vast crowd of stars is immediately presented to view. Many of them are rather large and quite bright, while the number of the smaller ones is quite beyond comprehension."

Even before Galileo's day, astronomers had known of other shining clouds in the heavens, which they called nebulae. The largest of these clouds (as seen from Earth) is the Andromeda Nebula, visible to the naked eye (if you are lucky enough to observe the night sky in conditions of extreme darkness, away from artificial lights) as a faint patch of light in the constellation Andromeda (hence the name). The telescope soon revealed many more of these patches of light in the sky, and Galileo suggested that they were simply clouds of stars that were too far away for even a telescope to resolve them into individual stars. The idea was taken up by a few people, notably Immanuel Kant, who in the mid-1750s picked up and publicised a suggestion made by the British astronomer Thomas Wright, that at least some of these nebulae might be "island universes," star systems like our Galaxy but far beyond the Milky Way.

But this was not a widely held view. Others suggested that these clouds really were clouds of material in space, and did not consist of large numbers of stars grouped together. By and large, though, for well over a hundred years after the invention of the astronomical telescope the nebulae were treated as an irritation, something to be avoided by astronomers looking for much more interesting things, like comets. Indeed, the first substantial catalogue of the positions of nebulae on the sky, containing just over a hundred objects, was compiled by the astronomer Charles Messier, in the 1780s, in order to prevent people who were unfamiliar with these objects "discovering" them and misidentifying them as comets.

Eventually, it turned out that there were different kinds of nebulae. Some are indeed clouds of gas and dust in space, part of the Milky Way Galaxy, and these have little relevance to the story of the cosmic distance scale. But some are indeed star systems — some are globular clusters, and the Andromeda Nebula, for example, is a galaxy rather like our own Milky Way Galaxy. These are at the heart of our current story. Some idea of how long the confusion in the study of these objects lasted, though, can be seen from the contributions of William Herschel at the end of the eighteenth century and the beginning of the nineteenth century. Herschel was a superb telescope maker, and, with his sister Caroline, carried out detailed surveys of the heavens, discovering the planet Uranus in 1781. Among other things, he studied many of the nebulae catalogued by Messier, and in 1784 he reported that he could see stars in twenty-nine of these nebulae, suggesting that it would only be a matter of time (and bigger telescopes) before all of these objects would be resolved into stars, the way Galileo's little telescope had resolved the Milky Way into stars.

But when Herschel looked at more of the objects on the list, even with bigger and better telescopes he found that they could not be resolved in this way. As his reflecting telescopes went up in size from 19 inches (48 cm) in diameter to 48 inches (122 cm), he found that many of the nebulae were indeed gas clouds, some of them surrounding individual stars. This tilted his views so far towards the other extreme that in 1811 he wrote: "We may also have surmised nebulae to be no other than clusters of stars disguised by their very great distance, but a longer experience and better acquaintance with the nature of nebulae, will not allow a general admission of such a principle."

A new factor was added to the confused picture in the middle of the nineteenth century, when the Earl of Rosse, an Irish nobleman who had both a keen interest in astronomy and the money needed to indulge that interest, built a huge telescope, with a 72-inch-diameter (183 cm) mirror, and discovered that some of the nebulae had a spiral pattern, like a whirlpool seen from above. Soon afterwards, in 1864, William Huggins made the crucial step of combining spectroscopy with astronomy, splitting up the light from astronomical objects gathered by a telescope and analysing the lines revealed in the spectrum. This showed, once and for all, that objects such as globular clusters were indeed collections of stars — their spectra were like the spectra of many stars added together. It also showed that other objects, such as the Orion Nebula, are made of hot clouds of diffuse gas, not thousands of stars crowded together. Unfortunately, though, the technique was not, then, good enough to reveal the true nature of even the Andromeda Nebula, let alone the fainter spirals investigated by Rosse. The light from these objects was simply too faint for the telescopes of the time to gather enough in for the spectroscopists to analyse.

Even the development of astronomical photography in the second half of the nineteenth century did not settle the issue, and at the beginning of the twentieth century the most widely held view was that our Milky Way represented the entire Universe (as we would now call it), that globular clusters were star systems much smaller than the Milky Way, but part of it, and that the faint nebulae were clouds of gas within the Milky Way Galaxy. But at that time, nobody even knew how big the Milky Way itself was.

Because the Milky Way itself is a band of light that stretches right around the sky, people like Thomas Wright and Immanuel Kant had reasoned, back in the eighteenth century, that we live somewhere in the middle of a flat, disc-shaped assemblage of stars. In the 1780s, William Herschel tried to put this guess on a more scientific footing, by counting the number of stars in 683 regions around the Milky Way, and found that there are roughly the same number of stars visible in patches of sky the same size everywhere around the Milky Way; more evidence, it seemed, that we live in the middle of what was then the known Universe. It was only in the middle of the nineteenth century, though, that astronomers began to determine accurate distances to stars, and in the early decades of the twentieth century the Dutch astronomer Jacobus Kapteyn updated Herschel's analysis by doing his own star counts and this time including estimates of the distances to the stars he was counting. He inferred that the Milky Way is a lens-shaped system (a disc with tapering edges, like a discus), about 10 kiloparsecs in diameter and 2 kiloparsecs thick, with the Sun some-where near the centre. We now know that Kapteyn was wrong, be-cause he had made no allowance for the dusty material between the stars (he couldn't; it hadn't been discovered then), which is concen-trated in the plane of the Milky Way, and acts like a fog in space, caus-ing so much extinction in starlight that it limits the distance we can see in the plane of the Milky Way. Kapteyn's "universe" was actually just the local slice of our side of the Milky Way Galaxy. But his numbers give you a feel for how small astronomers thought the entire Universe to be as recently as the second decade of the twentieth century.

The next big step out into the Universe came with the aid of those Cepheid variables that I have mentioned. But since their value as distance indicators was itself only discovered in the second decade of the twentieth century, it came too late to have an impact on Kapteyn's own work (he had been born in 1851, and died in 1922). The step built from the investigations of Henrietta Swan Leavitt, who worked at the Harvard College Observatory, under the supervision of the astronomer Edward Pickering. Pickering was engaged in an epic task of analysing and cataloguing thousands of stars. Many of these were stars from the southern skies, photographed by Pickering's brother William from an observing station in Peru — the southern hemisphere was a particularly fertile region for this kind of work at that time, because for obvious historical reasons most previous astronomical observations had been concentrated in the northern hemisphere.

One of the most interesting features of the southern sky is the pair of nebulae known as the Large and Small Magellanic Clouds, named after the explorer who was the first European to draw attention to them. By the end of the nineteenth century there was no doubt that these are star systems — they look like pieces of the Milky Way that have been broken off — but nobody knew the distance to them. As a matter of routine, Leavitt was set the task of identifying variable stars in these "clouds," by comparing photographic plates taken on different occasions to see if any stars had changed brightness in the interval. What she found came as a complete surprise.

The class of variables known as Cepheids are named after the archetype, Delta Cephei (the name simply means it is the fourth brightest star in the constellation Cepheus), which was studied by the English astronomer John Goodricke in the 1780s. They were already interesting to astronomers because although different Cepheids have different periods of variation, each individual Cepheid changes brightness in a very regular way, brightening and dimming before brightening again with a rhythm that repeats exactly, cycle after cycle. Some have cycles less than two days long, some about a hundred days long; but they all show the same pattern of behaviour (incidentally, most people have seen a Cepheid without realising it — the northern pole star, Polaris, is a Cepheid with a period of four days, but a very small range of brightness variation, undetectable to the human eye).

Leavitt found nearly two thousand variable stars in the Small Magellanic Cloud (SMC), and concentrated her attention on the ones with regular brightness variations, most of which (hundreds of individual stars) turned out to be Cepheids. But as the data came in, she began to realise that there was something special about these Cepheids. As early as 1908, she reported that the brighter Cepheids in the SMC had longer periods than the fainter Cepheids (that is, they go through their cycle more slowly). And in 1912 she published the discovery of a precise mathematical relationship between the brightness of a Cepheid in the SMC and its period. If the time it takes for a Cepheid to go through its cycle once is about eleven hours, for example, then that Cepheid is, on average, only one tenth as bright as a comparable star which has a period of about five days.

The brightnesses used in her calculations were, of course, the brightnesses as seen from Earth — the *apparent* magnitudes. But the inference was clear. All the stars in the SMC must be at roughly the same distance away from us, so that the loss in brightness by the light from any star in the SMC on its way to us must be the same (that is, all the stars have the same distance modulus). So what Leavitt had actually discovered was a relationship between the period of a Cepheid and its absolute magnitude — a period–luminosity relationship. This had never shown up in studies of Cepheids closer to home, because although one Cepheid might be twice as bright as another, it might also be twice as far away, masking brightness relationship. Now, though, if astronomers could measure accurately the distance to just one local Cepheid, by any of the standard techniques, they could determine its absolute brightness, and calibrate the period–luminosity relationship. By measuring its period, they would know where it belonged in the mathematical relationship that Leavitt had discovered. And that would mean they could turn the relationship around and use the measured period of any Cepheid in the SMC (or anywhere else) to work out its *absolute* magnitude. As ever, if you know absolute magnitudes *and* apparent magnitudes for the same objects, you know their distances (if you can allow accurately for extinction).

Astronomers were quick to respond to the discovery. In 1913, the Dane Ejnar Hertzsprung (the same Hertzsprung who was one of the inventors of the H–R diagram) applied the statistical parallax method to a sample of thirteen Cepheids in our neighbourhood, using this to find an "average" distance and brightness, which he converted into an absolute magnitude for a hypothetical Cepheid with a period of exactly one day. Using the measured periods of the SMC Cepheids studied by Leavitt, this gave him a distance to the SMC of 10,000 parsecs (more than 30,000 light-years). It was, in fact, an overestimate — he had made no allowance for extinction, which makes the distant stars look fainter. But it indicated the power of the new technique. It is worth putting this in perspective. It was only in the 1840s that the distances to a few stars were measured for the first time — distances of a few light-years. Now, almost exactly the "three score years and ten" of a human lifetime later (the lifespan of Pickering himself, who was born in 1846 and died in 1919, neatly makes the point), astronomers were talking about distances to astronomical objects ten thousand times bigger than the distances to those stars, which had seemed so remote in the 1840s.

In 1914, a year after Hertzsprung's contribution, in the United States Henry Norris Russell (the same Russell who was the other inventor of the H–R diagram) and his student Harlow Shapley carried out a similar analysis, making some allowance for interstellar absorption. There were problems with that piece of work, too; but it is worth mentioning because it marks the first appearance of Shapley in the story, and it was Shapley who would soon change astronomers' views on our place in the Galaxy, if not in the Universe. The best modern estimates of the absolute magnitudes of Cepheids, by the way, imply that one in the middle of the range of Cepheid brightnesses is ten thousand times brighter than the Sun, and the brightest are, in round terms, a thousand times brighter than RR Lyrae stars.

In 1918, Shapley went back to the Cepheid calibration, and came up with a revised baseline measurement, which gave him a reasonable yardstick to use to probe the size and shape of the Galaxy. But the new insight he gained from Cepheids came when he applied the period–luminosity relationship to variable stars in globular clusters, and worked out their distances from us. There was an element of luck in this — the variable stars that Shapley picked on to study in globular clusters were, we now know, RR Lyrae stars, not Cepheids. Because RR Lyrae stars are fainter than Cepheids, this meant that the distances Shapley worked out were too big — what looked to him like a bright Cepheid far away was actually a dimmer RR Lyrae star, not quite so far away. On the other hand, he still wasn't making proper allowance for extinction, and if he had he would have placed the stars he was studying much closer (a bright, nearby star suffering extinction looks like a fainter, more distant star with no extinction). To some extent, the two errors cancel each other out, so the distances Shapley ended up with were in the right sort of ballpark. And, as it happened, the exact distances to the globular clusters were not needed for Shapley's most important discovery to be apparent. Simply from the relative distances to the clusters (knowing that cluster A is twice as far away as cluster B, and so on) Shapley could work out how the globular clusters were distributed relative to the Sun. He found that they are not distributed in a sphere around the Sun; rather, they are distributed in a sphere (like plums in a plum pudding) around a point thousands of parsecs away from us, in the direction of the constellation Sagittarius, which is in the middle of the band of light formed on the sky by the Milky Way.

This was apparent, in spite of the large amount of extinction that (we now know) plagues observations in the plane of the Milky Way, because many of the globular clusters are seen high above (or deep below) the plane of the Milky Way. This is rather like the way tall skyscrapers in the centre of a city can be seen from the suburbs. You cannot see the base of the skyscraper, because all the houses at ground level get in the way of your line of sight (extinction). But by looking upwards, you can see the tops of the skyscrapers. If you had a way to measure the distances to the tops of the individual skyscrapers (in this case, radar would do), you could map the city centre without ever leaving your home. Shapley was actually seeing some globular clusters above and below the plane of the Milky Way on our side of the Galaxy, some more or less at the distance of the centre, but high above (or below) it, and some more far beyond the centre of the Milky Way, above or below the plane of the Galaxy on the far side from the Sun.

The implication of this discovery is that we do not live in the middle of the Milky Way, and also that the whole Milky Way Galaxy is much bigger than had previously been thought. Instead of a Milky Way about 6,000 parsecs in diameter centred on the Sun, in 1920 Shapley described a Milky Way about 100,000 parsecs across, with the centre of the Milky Way about 10,000 parsecs away from us. This was rather too big, but for the first time astronomers had an indication of the relative place of the Sun and Solar System in the Milky Way. Modern estimates give a diameter of about 28,000 parsecs, with the Sun some 8,000 or 9,000 parsecs from the centre of the disc, which is itself only a couple of hundred parsecs thick (really *very* thin, compared with a diameter of 28 *thousand* parsecs). The globular clusters are distributed in a sphere surrounding the whole thing, centred on the centre of the Milky Way. And the important point is that these numbers are more or less in the same ratios as the ones found by Shapley, even though they are all smaller.

The unfortunate thing about Shapley's estimate of the size of the Milky Way is that by making it so huge, it made it easier for astronomers (and Shapley in particular) to accept the idea that all of the nebulae were systems that belonged in our Galaxy, or at most might be smaller star systems around the edges of the Milky Way, like islands off the coast of a great continent.

But not everyone agreed with Shapley's interpretation of the evidence. The Cepheid distance scale was still such a new idea, and based on such a small sample of stars, that other astronomers who didn't like Shapley's conclusions still felt able to dismiss it as unreliable, and to throw out all of those conclusions based on Cepheid observations. Those astronomers who favoured the idea that some of the nebulae — in particular, the spirals — might be galaxies like the Milky Way were inclined to believe that the Milky Way must be much smaller than Shapley suggested. One of the leading proponents of this idea in the second decade of the twentieth century was the American astronomer

Heber Curtis, who became an expert at photographing spiral nebulae and analysing their appearance. Among other things, he noticed that in photographs of thin nebulae, interpreted as spirals seen edge-on, there was always a dark line along the middle of the nebula, suggesting an accumulation of dusty debris in the plane of the disk. If a similar accumulation of obscuring material existed in our Milky Way, it would help to explain many of the puzzles of extinction, and suggest a close similarity between the Milky Way and spiral nebulae.

The issue was considered so important that in 1920 the US National Academy of Sciences arranged a head-to-head debate between Shapley and Curtis. Shapley argued that the Milky Way Galaxy is about 100,000 parsecs across, that our location in it is far away from the centre, and that it is essentially the entire Universe. Curtis argued that the Milky Way Galaxy is only about 10,000 parsecs across (perhaps going to extremes in his efforts to distance himself from Shapley), that it is just one "island universe" among many, probably a spiral nebula, and that the Sun is near the centre of the Milky Way.

The debate was inconclusive, as it had to be since each side was partly right and partly wrong (both sides claimed victory, which is a sure sign that it was inconclusive). From our point of view, the most interesting aspect of the debate concerns the nature of the nebulae, and the final acceptance of Cepheids as reliable distance indicators. It was resolved within five years of the NAS meeting in 1920, and it was resolved, as these things usually are, by new and improved observations. But before we move on to the story of the nebulae, I want you to be quite clear about how much we depend upon a few observations of relatively nearby stars. In his book *The Cosmological Distance Ladder*, published as recently as 1985, Michael Rowan-Robinson, of the University of London, surveyed all of the distance-measuring techniques known to astronomers. He could find only twenty Cepheids with reliably determined distances that could be used as calibrators, and even for two of these he listed some of the brightness measurements as "uncertain." Other astronomers have used distances to other Cepheids, determined by various less accurate techniques, in their calculations; but these eighteen are the only ones for which statistical parallax works. The number is scarcely any bigger than the number used by Hertzsprung (thirteen), for the very good reason that any Cepheid close enough to be used as a calibrator is bound to be bright enough, as seen from Earth, to have been noticed by astronomers at the beginning of the twentieth century. There simply are no more Cepheids close enough to do the trick.

I'm not saying that there is anything seriously wrong with the Cepheid calibration — it looks better than ever today, thanks to improved observations and comparison with computer models. But as we take the first step out into the Universe at large, it is worth remembering how lucky we are to have any kind of yardstick at all with which to measure the scale of the Universe.

4

Into the Blue

Beyond the Milky Way

It wasn't only his overestimate of the size of the Milky Way (based, we now know, on an incorrect calibration of the Cepheid distance scale) that led Shapley, and others in the early 1920s, to think that the other nebulae, even the spirals, were lesser satellites of our own Galaxy, or even part of the Milky Way itself. There were two particularly troubling things that the Curtis camp had to come to terms with — both of them, we now know, wrongly interpreted at the time, but there was no way for Curtis to know that.

The first puzzle was the appearance of a bright star that had suddenly flared up in the nebula known as M31 (the Andromeda Nebula) back in 1885. The spiral nature of this nebula was discovered by the astronomer Isaac Roberts in the same decade (we now know that M31 is the nearest spiral galaxy to the Milky Way), so it was regarded as a classic test case in determining the nature of these objects. And the star that flared up in the nebula in 1885 was photographed, so there was ample opportunity for succeeding generations of astronomers to study the phenomenon, and no room to doubt that it really had been as bright as contemporary observers had reported. The problem — for those who argued that spiral nebulae were independent systems on the same of the Milky Way — was that this "new" star was simply too bright.

Astronomers had already seen (and even photographed) stars that flare up in this way in the Milky Way itself — they are called novae, from the Latin word for new, although really they are old stars that flare up into short-lived prominence, not literally new stars. The apparent brightness of the nova seen in M31 in 1885 was more or less the same as the apparent brightness of a typical nova seen in the Milky Way itself. If all novae have roughly the same absolute magnitude (which seemed a reasonable guess at the time), that meant that the Andromeda Nebula must be a cloud of gas associated with one or more stars somewhere in the Milky Way itself. And if one spiral nebula was, in fact, part of the Milky Way, probably they all were.

The argument can be turned on its head. If the Andromeda Nebula really is a galaxy like our Milky Way, we can make another "reasonable guess." At a rough guess, it ought to be about the same size as the Milky Way. This immediately brings us back to traditional surveying techniques. Remember the standard rod used by surveyors? If we guess that the Andromeda Nebula is the same size as the Milky Way, we can use it as a standard rod for triangulation, in exactly the same way that the surveyor measures the width of the river without getting wet feet, by sending the assistant over to the other side with the standard length rod. If we know (or guess) the actual size of the Andromeda Nebula (its linear size), then the apparent size of the nebula on the sky (its angular size) tells us how far away it is, by triangulation. Of course, we don't know the size of the nebula; but if it is roughly the same size as the Milky Way itself then it must be at a huge distance in order to appear as a small patch of light on the sky. And if it were that far away, then the nova seen in 1885 would have to be spectacularly bright, far brighter than any nova that nineteenth-century astronomers had seen in the Milky Way. It was straightforward to calculate that if the Andromeda Nebula were as far away as Curtis suggested, then the nova of 1885 would have been at least as bright as a billion stars like the Sun put together. This seemed impossible at the time; but we now know that very rare stellar outbursts do occur, in which a single star briefly shines not merely as brightly as a billion suns, but as brightly as a hundred billion suns. They are called supernovae, and they have their own part to play in the story of the investigation of the age of the Universe, which we will come on to later.

The point of telling the story in such detail, though, is not to say who was "right" and who was "wrong" in the debate in 1920. What I want to emphasise is that "reasonable assumptions" can be wildly misleading, and that different reasonable assumptions about the same thing can lead to diametrically opposite conclusions. Only actual observations of the real Universe (and, where possible, comparison with experiments carried out in the laboratory, or computer models) can tell us which explanation of a cosmic phenomenon gives a more accurate portrayal of what is going on.

In fact, even by the time of the Curtis–Shapley debate there were hints that there was something unusual about the "nova" seen in M31 in 1885, because fainter outbursts (now known to be ordinary novae) had been seen and photographed in several nebulae by then, and applying the same rule of thumb to them, that they probably had absolute brightnesses similar to the absolute magnitudes of novae in the Milky Way, did give distances to those nebulae far beyond the borders of the Milky Way.

But there was still no proof, and Shapley had another seemingly powerful reason to think that the spiral nebulae were nearby and probably part of the Milky Way system. His reasons for thinking so, though, highlight another problem in all of science, not just astronomy — wishful thinking.

The problem was that the Dutch astronomer Adriaan van Maanen thought that he had measured the rotation of several spiral nebulae. The way he tried to measure the rotation was straightforward enough — take photographic plates of the same objects at intervals several years apart, and look to see if identifiable features in those nebulae have moved round in the interval (much the same as the way you can measure proper motions of stars across the sky by comparing photographs of the same part of the sky taken years, or decades, apart). As early as 1916, van Maanen had claimed that he could detect a tiny annual rotation of the nebula known as M101, amounting to 0.02 of a second of arc. If these measurements were correct, they meant that M101 had to be relatively nearby, because the angular rotation can be translated into a linear speed corresponding to the distance to the object. At the kind of distances required by Curtis, the kind of rotation reported by van Maanen meant that M101 would be rotating faster than the speed of light. By the early 1920s, van Maanen had half a dozen more spirals on his list of objects with detected rotation rates, all pointing to the same conclusion.

Everyone agreed that if he was right these measurements were fatal to the idea of spiral nebulae as independent galaxies; but most people had serious reservations about accepting the measurements at face value. It wasn't that they thought van Maanen was making it up; just that the measurements were so incredibly difficult that it was hard to believe that he was seeing what he thought he was seeing. After all, the sizes of the effects van Maanen thought he was seeing are equivalent to about 0.001 percent of the angular diameter of the Moon as seen from Earth. But — and this is the point — van Maanen was a friend of Shapley. Shapley trusted him, because he was a friend; so he took the reports of measured rotation rates of spirals at face value. Many more observations over the past seven decades have, however, shown that van Maanen was just plain wrong.

Richard Feynman used to say that, because of wishful thinking, the easiest person to fool in science is yourself. The sad moral of this particular tale, though, seems to be that the easiest person to fool is your friend. The moral is twofold — don't accept evidence just because it comes from a friend, or because it lends weight to your own pet theory; but equally don't reject evidence just because it comes from someone you dislike, or because it destroys your cherished theory. Either way, double-check the evidence and, for better or worse, accept the results of a cold-blooded appraisal of its merits.

Whatever the reasons, though, the confusion about the nature of the spiral nebulae persisted in the first half of the 1920s, and could only be resolved by making direct measurements of the distances to at least some of the spirals. That meant finding Cepheids in them; and that required new technology, bigger and better telescopes than anything that had been available before. Everything came together in a scientific paper read to a joint meeting of the American Astronomical Society and the American Association for the Advancement of Science held in Washington, D.C., on 1 January 1925. But by then, both Curtis and Shapley had left the arena.

Curtis had been the first to move on. He was forty-eight in 1920, the year of the debate with Shapley, and had been working at the Lick Observatory on Mount Hamilton, near San Jose in northern California. The same year, he took up a post as Director of the Allegheny Observatory, and became primarily an administrator, making little further direct contribution to astronomical research. Shapley was much younger than Curtis (he had been born in 1885), and had been working at the Mount Wilson Observatory, near Pasadena, in southern California (the proximity of the two observatories lent spice to the friendly rivalry between Curtis and Shapley). He moved on in 1921, and with hindsight this seems to have been a singularly ill-judged career move — even though he left Mount Wilson to become the Director of the Harvard Observatory, a post he held until 1952, building the reputation of the observatory and contributing to the training of a succession of major names in astronomy. Back on Mount Wilson, though, at the observatory Shapley left behind it was a new telescope, in the hands of two of the greatest observers of the twentieth century, that first settled the question of the nature of the spiral nebulae and then revealed that the Universe had a definite beginning in time.

The Mount Wilson Observatory had been built around a telescope with a 60-inch (152 cm) reflecting mirror, which came into operation in 1908. Just ten years later, this was joined on the mountain by the 100-inch (254-cm) Hooker Telescope (named after the benefactor who paid for it), which was to be the most powerful astronomical telescope on Earth for nearly thirty years, until the completion of the famous 200-inch (508-cm) Hale Telescope (named after George Ellery Hale, the astronomer who created both the Mount Wilson and the Mount Palomar observatories), at Mount Palomar, near Los Angeles (not far from Pasadena), in 1947. And the two people who would push the 100-inch to its limits in the 1920s were already working on the mountain before Shapley left.

The first of those pioneers, Milton Humason, became an astronomer in such an unusual way that it is well worth digressing a little to give you some details of his background. Humason was born in

Dodge Center, Minnesota, on 19 August 1891; but his parents moved the family to the West Coast when he was a child. At the age of fourteen, in 1904, Humason was taken to a summer camp on Mount Wilson (this was about the time the observatory was being established), and fell in love with the mountain. He persuaded his parents to let him take a year out from his education, and got a job at the then-new Mount Wilson Hotel (lower down the mountain than the observatory), working as a bellboy and general handyman, and looking after the pack animals which were used in those days to carry goods (and people) up the mountain trails.

Humason never went back to school. Instead, by the end of the decade he became a mule driver, working with pack trains carrying equipment right up to the peak of the mountain, where the 60-inch reflector (then the best astronomical telescope in the world) had become operational, and work was in progress on the dome and other buildings associated with the planned 100-inch telescope. Every item of equipment for the observatory, from the telescopes themselves, to lumber and other building supplies, and the food for the construction gangs and the astronomers, went up the mountain this way. This is as good an indication as any of just how much technology has changed since the early twentieth century, and just what an achievement the 100-inch was in its day. There was also the minor point that anyone working on the mountain had to keep a careful lookout for mountain lions, which still roamed the peak then.

While working on the mule trains and enjoying the outdoor life, Humason fell in love with Helen Dowd, the daughter of the engineer in charge of the activities on the mountain peak, and the couple were married in 1911, when they were both just twenty years old. The arrival of a baby, William, in the autumn of 1913 persuaded Milton that he ought at last to think about putting down some roots, and for three years he worked as head gardener on an estate in Pasadena (some reports describe this as being "foreman on a ranch," but even in 1914 Pasadena wasn't exactly the Wild West; the term *ranch* was often used in the same way that we would use the word *farm*).

Three years later, the young couple purchased their own "citrus ranch"; but almost immediately an opportunity came up that Milton and Helen, who had been pining for the mountain, couldn't resist. Helen's father told them that one of the janitors at the observatory was about to leave, and suggested that the job might suit young Milton. Even better, with the 100-inch telescope due to become operational in 1918, there was a chance to combine the janitorial duties with the post of "relief night assistant," helping out the astronomers, if required, on both the big telescopes. The pay was modest — eighty dollars per month — but the post included rent-free accommodation, and free meals while working. And it meant living on the mountain (by all accounts, if he had had any money Humason would have paid *them* to let him live on the mountain). He took up the post in November 1917.

Within a year, Humason had learned how to take photographic plates of astronomical objects, using the smaller telescopes on the mountain, and he proved so adept at this arcane art that in 1920 he was officially appointed to the astronomical staff of the observatory, largely at Shapley's recommendation (George Ellery Hale was still the Director of the Mount Wilson Observatory at that time, and made the appointment, we are told, rather grudgingly). There were some mutterings about this promotion of the high school dropout and mule skinner, who just happened to be the son-in-law of the observatory's chief engineer; but these were soon stilled as Humason's remarkable ability at obtaining astronomical photographs became clear.

Shapley described Humason as "one of the best observers we ever had"; from a distance of nearly eighty years, he looks like *the* best observer on the mountain in the 1920s and 1930s. And this was quite an achievement. The arcane skills involved in getting images of faint astronomical objects in those days began with the actual observations. This meant sitting at the telescope night after night (perhaps every night for a week), keeping it pointing accurately at the object of interest (typically, a galaxy, in Humason's case) while the light from the object was gathered in and directed to a glass photographic plate (coated with light-sensitive material) at the focus of the telescope. In those pre-computer days, the telescope needed constant human attention to keep tracking perfectly across the sky to compensate for the rotation of the Earth and hold the same celestial object centred in its field of view for hours on end — it did have an automatic tracking system (essentially a clockwork mechanism controlling electric motors), but this had its own little foibles and could not be left unattended. And, of course, the dome had to be open to the sky, so the telescope could see out, and it had to be unheated, because convection currents of air rising past the telescope would blur its field of view. Even in summer, the mountain top can be cold at night (I visited it in May one year, when there was snow on the ground); and the best time to observe, of course, is in the depths of winter, when the skies are dark for longest. One other thing — there could be no artificial light inside the dome, apart from a dim red bulb, because that would fog the photographic plates.

Each night, working under these difficult conditions, the same plate would be carefully exposed to the light from the telescope at the start of the observing run, and carefully shut away in a dark container at the end of the night's observing. Only after a week or so would enough light have been gathered to provide a good image of the object. Then, the observer would have to process the plate, by hand, in the dark (a fragile *glass* plate, remember), using a variety of chemicals first to develop the picture, and then to fix it as a permanent image on the plate. Use the wrong strength of chemicals, or apply them for the wrong amount of time (or let the plate slip from your grasp), and a week's work would be ruined. Extreme patience and a calm, unflappable manner were essential requirements for a successful observational astronomer in those days — as it happens, characteristics that are also required of a successful mule driver.

Even though he became the best observer on Mount Wilson, and probably the best in the world, Humason was always diffident about his lack of academic qualifications, and understandably cautious (especially in his early years as an astronomer) about pushing his own ideas forward. The combination of his great skill at astronomical photography and this understandable diffidence led to a bizarre incident, which happened shortly before Shapley left the mountain to take up his post at Harvard. This was early in 1921, the year after Humason had been appointed to the astronomical staff in the most junior capacity, and the year before he received the dizzying promotion to the rank of assistant astronomer. Humason had been given the task (by Shapley) of comparing plates of the Andromeda Nebula, M31, obtained by the new 100-inch telescope on different occasions, to see if there were any differences in the images (this was probably in an attempt by Shapley to find evidence for the kind of rotation that had been claimed by van Maanen for other nebulae). The way this kind of comparison was made (and still is, on some occasions, although computers have largely taken over) was to "blink" the plates in a special kind of viewer. Looking through the eyepiece of this device, you see each plate in turn, repeatedly, with the image bouncing backward and forward between the two. When this is done, any differences in the two images leap out at the human eye.

To Humason's surprise, when he blinked the plates of the Andromeda Nebula in this way, he thought he could see tiny specks of light that were present on some plates but not on others—as if there were variable stars in the nebula. He carefully took the plate with the best example of this, and marked the positions of the interesting features with little lines drawn in ink on the back of the plate. Then, he took the plate to show Shapley what he had found. Shapley simply ignored Humason's claim. First, he explained to the most junior astronomer on the mountain just why it was impossible for there to be variable stars in the Andromeda Nebula, essentially going over the same arguments that he had used in the debate with Curtis. Then, he took a clean handkerchief out of his pocket, turned the plate Humason had given him over, and wiped away the identifying ink marks. A few weeks later, on 15 March 1921, he left for Harvard.

Humason said nothing to anyone at the time, for obvious reasons. He had barely got his foot on the first rung of the astronomical ladder, and he owed even that modest position largely to Shapley's recommendation. But later in his career he told the tale on several occasions, and one of the interested listeners was Allan Sandage, who will have a big part to play in our story later on, and from whom this version of the story comes. There are many tantalising "what ifs" hanging around the story. If Shapley had stayed at the Mount Wilson Observatory, might he have had second thoughts, and discovered the truth about the spiral nebulae? Or would his stubbornness have had an influence on his colleagues there, and held back the discovery of this truth? It is a fruitless game to play, but once again the moral of the story is clear—you have to accept the observations (or at least, take them seriously enough to look at them in detail), even if they conflict with your cherished theory.

The other pioneer who, together with Humason, reshaped our

understanding of the Universe in the 1920s, took that attitude to extremes. Edwin Hubble never really subscribed to any theory about the Universe at all, in spite of the association made today between his name and the theory of the Big Bang. Hubble was an observer, and he reported the observations he made almost entirely without any trappings of theoretical interpretation, leaving that for others to do. He also came from a background of academic achievement that contrasts sharply with the background of Humason, with whom his name will always be linked — although, as we shall see, Hubble always exaggerated his own social status and achievements outside astronomy.

Hubble had been born in Marshfield, Missouri, in 1889. He was one of eight children; their father, a failed lawyer, worked in insurance and travelled widely as a manager overseeing scattered offices, so as a child Edwin's adult male role models were his two grandfathers. It was his maternal grandfather, a medical doctor called William James, who, we are told, introduced Edwin to the wonders of astronomy by building his own telescope and allowing the young boy to look through it at the stars as a treat on his eighth birthday.

At the end of 1899, the family moved to Evanston, Illinois, on the shore of Lake Michigan, and in 1901 to the newly incorporated city of Wheaton, just outside Chicago. So it was in Chicago that Edwin Hubble attended both high school and university, making a name as a good athlete (although not quite the all-round star that he would later lead people to believe) and as a first-class scholar. After studying science and mathematics for two years, and being awarded the two-year Associate in Science Degree, Hubble concentrated on courses in French, the Classics, and political economics, aiming for a Rhodes Scholarship, which he duly won. He received his Bachelor's degree in 1910, then took up the Scholarship at Queen's College, in Oxford, where he studied law and acquired an exaggerated "Oxford Britishness" in speech and mannerisms that stayed with him for the rest of his life.

Hubble's father died in 1913, at the early age of fifty-two, a few months before the Rhodes Scholar returned from England. During what must have been a traumatic year that followed, Edwin helped to settle his father's modest estate and made sure that the family, now living in Louisville, stayed together. In spite of his later claims to the contrary, he never practised law, but he did work for a year as a high school teacher. His immediate duty by his family done, in 1914 Hubble moved on to the Yerkes Observatory (part of the University of Chicago) as a research student in astronomy (it is perhaps worth mentioning that he was only able to do this because his younger brother, Bill, largely took over the financial responsibility of looking after Hubble's mother and sisters).

The Yerkes Observatory was the first to have been founded by Hale (who by 1914 had long since moved on to Mount Wilson), using funds provided by the millionaire Charles T. Yerkes, who made his money out of trolley cars. The main instrument there was a 40-inch (102-cm) refracting telescope (one that uses lenses, not mirrors), which was then one of the best astronomical telescopes in the world, and is still the largest refractor ever made (and still in use). Hubble's main work as a student and research assistant between 1914 and 1917 was to photograph as many of the faint nebulae as possible — by the time he joined the observatory, about seventeen thousand nebulae had been catalogued, and it was estimated that perhaps ten times more might be visible, in principle, to the 40-inch at Yerkes, the new 60-inch reflector on Mount Wilson, and comparable instruments. But this was still, remember, before the distinction between nebulae that are part of the Milky Way and what we would now call galaxies was recognised. Hubble's first contribution to astronomy was an attempt to classify the nebulae according to their appearance; but although his work was good enough for him to be awarded his Ph.D. in 1917, little came of these efforts for another five years, partly because of America's involvement in World War One.

Even before he completed his Ph.D., Hubble had been offered a post on Mount Wilson by Hale, who had been head-hunting to carry out an increase in staff on the mountain in anticipation of the 100-inch telescope becoming operational, and naturally turned to Yerkes as a source of suitable candidates. In fact, Hubble had wanted to stay at Yerkes, but there were no funds available for him there, so he had little choice but to accept the offer from California. But in April 1917, the United States declared war on Germany, in response to the German policy of unrestricted submarine warfare. Hubble volunteered for the infantry as soon as he had completed the formalities for his Ph.D., and Hale promised to keep the job at Mount Wilson open for him until he returned from Europe.

Hubble's own account of his military experiences differs from the official records, although there is no doubt that he achieved the rank of Major. His division, the Eighty-Sixth, reached France only in the last weeks before hostilities ended, and never saw combat. Yet Hubble always said (or implied) that he had been in action and had been wounded by shell fragments, which was why he could not straighten his right elbow properly. He also managed to linger in England, which he loved, for long enough before returning to the United States for an irritated Hale to write urging him to make haste, since the 100-inch was operational and there was plenty of work to do. But it was not until 3 September 1919 that Major Hubble (he liked to use the title even in civilian life) finally joined the staff of the observatory on Mount Wilson, when he was only a couple of months short of his thirtieth birthday.

Hubble first made his name as an astronomer by developing the ideas from his doctoral thesis, and coming up with a classification scheme for galaxies (I shall use the modern term, although Hubble always preferred the word *nebulae*). One of the important early contributions made by Hubble was the recognition that there are huge numbers of another kind of object, different from the spiral nebulae, which also seemed impossible to explain in terms of phenomena contained within the Milky Way. These are now known as elliptical galaxies. The differences between ellipticals and spirals are completely unimportant for the purposes of the present book; all that matters is that in due course it was realised that both kinds of nebula are indeed galaxies in their own right. It is now thought that ellipticals (which range in appearance from spherical to a flattened convex lens shape, like the profile of an American football) are formed by mergers between spirals, explaining (among other things) why the largest galaxies known are ellipticals. But none of this was known to Hubble in the early 1920s. The classification scheme he developed was essentially complete by the summer of 1923, although it wasn't published until some time later. For once, there was a theoretical model attached to the scheme, although as it turned out the theory was wrong — nonetheless, the scheme has proved useful.

Hubble's idea was to place the galaxies in a sequence along a line, starting with the spherical galaxies and moving through the lens-shaped ellipticals to the spirals. At this point, his diagrammatical representation of galaxies forked into two branches. On one branch, tightly wound spiral galaxies gradually gave way to more loosely bound spirals; on the other, there was the same unwinding of the spiral structure, but the spiral galaxies were marked by a short bar of stars across the centre of the pattern, with the spiral arms twisting outward from the opposite ends of the bar. This "tuning fork" diagram originally implied an evolutionary sequence, with a galaxy starting out as spherical, becoming more elliptical as a result of rotation, developing spiral structure with or without a bar, and with the spiral pattern loosening up as the galaxy aged. This idea (which developed from a proposal by the astronomer James Jeans) was completely wrong; but the Hubble classification is still used as a convenient way of labelling galaxies, and all except for a relatively small number of irregularly shaped systems (like the Magellanic Clouds) fit into the classification somewhere.

While Hubble was gathering his evidence in favour of this classification scheme, and becoming increasingly adept at using the 100-inch, the debate about the nature of the nebulae had continued to flicker. The "island universe" idea had been championed by the Swedish astronomer Knut Lundmark in his doctoral thesis in 1920, and in 1921 and 1922 he visited both the Lick Observatory and Mount Wilson, obtaining spectra of the spiral known as M33, and convincing himself (but certainly not Shapley) that the speckled, grainy appearance of the nebula meant that it was indeed composed of large numbers of stars. In 1922, three variable stars were identified in the patch of sky covered by M33, but the observations of these very faint objects were not good enough for the nature of these stars to be determined; in 1923, a dozen variables were found in another nebula, NGC 6822, but again the observations were not good enough to identify the nature of these stars immediately (it took a year's observations before they were eventually identified as Cepheids, and by then this was no surprise).

The search for Cepheids in nebulae didn't look too promising in the middle of 1923, when Hubble had completed his work on the classification scheme, but the prospect of finding novae in the nebula, using the 100-inch, looked much more promising. If ordinary novae could be firmly identified in M31, proving that the object seen in 1885 was a rarer and much brighter outburst, that would be as good a way as any of establishing the approximate distance to the nebula.

It was with this in mind that Hubble began another observing run with the 100-inch in the autumn of 1923, concentrating on photographing one of the spiral arms in the Andromeda Nebula, M31. Seeing conditions were poor on the night of 4 October, but even so a forty-minute exposure produced a plate with a bright spot, possibly a nova. The next night, a slightly longer exposure confirmed the presence of the nova, and showed two more spots of light — two more suspected novae. Back in his office, Hubble dug out earlier plates showing the same part of M31, going back several years and obtained by various different observers, including Humason and (ironically) Shapley. It was this series of plates which, under close examination, showed that one of the two additional "novae" discovered by Hubble on 5 October was, in fact, a Cepheid variable with a period of just under 31.5 days. Plugging in the period–luminosity relationship and distance calibration used by Shapley himself in his survey of the Milky Way Galaxy, this immediately gave Hubble a distance of 300,000 parsecs to the Andromeda Nebula — almost a million light-years, and three times the size of what Shapley had considered to be the entire Universe. Since then, partly because of the calibration problems I have mentioned, the estimated distance to M31 has been revised up to about 700 kiloparsecs; but even with the incorrect calibration, Hubble had proved that at least one spiral nebula was indeed an object comparable in size to our Galaxy, and far beyond the Milky Way.

This distance estimate also explained why observers such as Lundmark had had such trouble finding convincing evidence of individual stars in M33 and other spiral nebulae. The smallest features identifiable on the best photographic plates of the heavens available at the time covered an angle on the sky of half a second of arc (more than 3,500 times smaller than the angular size of the Moon seen from Earth). But at a distance of a million light-years, even such a tiny angle corresponds to a region 2.5 light-years (nearly a parsec) across — more than half the distance from the Sun to Alpha Centauri. Any bright spot seen in those photographic plates could be a single star, or a group of stars, or a cloud of hot gas, as long as the whole conglomeration occupied a volume less than 2.5 light-years in diameter. As we shall see, this kind of problem caused recurring difficulties with estimates of the distance scale to galaxies, and hence the age of the Universe.

Of course, it would take more than one Cepheid to convince Shapley. Over the winter months of 1923–24, Hubble found nine novae and another Cepheid in M31, all pointing to the same conclusion. In 1924, he found nine Cepheids in another nebula, NGC 6822, fifteen in the spiral M33, and others in other nebulae. Even Shapley had now to admit defeat, and it was this body of work that formed the basis of the paper presented to the AAS/AAAS joint meeting on 1 January 1925 (Hubble was not present at the meeting, and the paper was read to the meeting by Henry Norris Russell on his behalf). The view of the meeting was that the question of the nature of these nebulae had finally been resolved, and that the Universe extended far beyond the confines of the Milky Way.

Hubble's place in the history books would have been assured even if he had given up astronomy on the spot. But there was another pressing puzzle about the nature of the nebulae, one which cried out for careful study using the best telescope on Earth, the 100-inch. It was a puzzle that had been building up for more than a dozen years, since Hubble was a Rhodes Scholar in Oxford, where he knew nothing of the work being carried out at the great observatories in the United States.

5

Hubble's Law

A Universe with a Beginning

Science, as we have mentioned, does not proceed in a neat, linear fashion with discoveries following one another and slotting into place. And in order to get a grip on what Hubble (and Humason) did next, we have to go back more than a dozen years from 1925, to the discovery, made by Vesto Slipher, of what seemed to be huge Doppler shifts in the light from many of the nebulae.

Once again, Mars comes into the story, although this time rather more tangentially (remember that our knowledge of distances across the Solar System, which define the astronomical unit and thereby the parsec, started with measuring the distance to Mars). In the late nineteenth century, there was a flurry of interest in the planet Mars among American astronomers, and the public, largely triggered by the work of the Italian Giovanni Schiaparelli. He had observed and described features on the surface of Mars, which he called *canelli*, meaning "channels." This word was mistranslated into English as "canals," and fired a wave of interest in the possibility of intelligent life on Mars — among other things, providing the inspiration for H. G. Wells to write *The War of the Worlds,* first published in 1898.

Percival Lowell, a wealthy American businessman (the family money came from cotton) with a lifelong interest in astronomy, became so intrigued by the idea that in 1894, at the age of thirty-eight, he decided to set up a private observatory at Flagstaff, Arizona (an excellent site, more than 2,000 metres above sea level). His primary purpose was to prove the existence of life on Mars, and although the observatory did carry out other work, this remained a major influence on what went on there until Lowell's death in 1916. Along the way, though, he built up a first-class observatory, equipped with fine telescopes, and did a great deal to popularise astronomy. The Lowell Observatory remains an important research centre today.

Lowell's interest in Mars extended to the other planets of the Solar System, and the puzzle of how such planetary systems form in the first place. At the beginning of the twentieth century, one of the possible explanations of spiral nebulae was that they were swirling clouds of gas and dust, within the Milky Way system, each settling down to form a central star surrounded by planets. So, naturally, Lowell was interested in spiral nebulae, and he set one of his small team of astronomers, Vesto Slipher, the task of investigating spirals in the hope that they would provide clues to the formation of the Solar System.

Slipher, who had been born in 1875, graduated from the University of Indiana in 1901, and joined the staff of the Lowell Observatory the same year. Some of his early work there led to the award of a Ph.D. (also from the University of Indiana) in 1909, and when Lowell died Slipher took over the running of the observatory, where he was Director until he retired in 1952. His work on spiral nebulae was right at the cutting edge of research at the beginning of the second decade of the twentieth century, comparable in its day to the development of new telescopes and electronic light detectors (charge coupled devices) today. Slipher used a very good telescope, a 24-inch (61-cm) refractor (by and large, a refractor is more powerful than a reflector with the same diameter), and a new kind of detector, which enabled him to measure the positions of lines in the spectra of at least the brighter spiral nebulae. He was also an immensely skilful and patient observer.

It's worth emphasising just how difficult this work was. Astronomical photography was only a few decades old; astronomical spectroscopy dated only from the end of the 1850s. And putting the two together posed extra difficulties. One of the problems with astronomical spectroscopy is that it involves spreading out the light from an object to make the spectrum, and using this to study the spectral lines. But astronomical objects are faint to start with, and when you spread the light out in this way (for example, using a prism) it is fainter still — often too faint to form a usable image on the kind of photographic plate available in the early days of this type of work. So, naturally, the first astronomical spectra were obtained for the Sun and the brighter stars (the first Doppler measurement of the velocity of a star was made in 1868, by William Huggins).

To put things in perspective, the breakthrough discovery of helium on the Sun by spectroscopy had been made only in 1868, just seven years before Slipher was born, and helium was only identified on Earth, finally confirming the power of astronomical spectroscopy, in 1895. The first emission spectra of spiral nebulae were actually obtained by Edward Fath, a graduate student at the Lick Observatory, who photographed the spectra of seven spirals and presented the data in his doctoral thesis in 1909. Fath provided what looks today like convincing evidence that the spirals are at least partly composed of stars, and could not just be clouds of gas, by comparing his spectra with spectra of stars — in a neat test of the idea, he deliberately put his telescope (a 36-inch [91-cm] reflector) slightly out of focus to get blurred images of three globular clusters, producing spectra similar to the ones he had obtained for the seven spirals. But the work made little impact and was largely ignored at the time (perhaps because it came "only" from a graduate student, not from a big-name astronomer).

Slipher was certainly pushing the technology to the limit in 1912 when he managed to obtain photographs of spectra (spectrographs) of the Andromeda Nebula, clearly showing lines in the spectrum of light from the nebula. Once he had done this, though, it was straight-forward to identify lines in the spectrum corresponding to known elements (a step which Fath might conceivably have carried out, if he had been a more experienced astronomer), and to measure their exact positions in the spectrum. The astonishing result of these measure-ments was the discovery that the lines were shifted significantly to-wards the blue end of the spectrum, indicating (by the Doppler ef-fect) that the Andromeda Nebula is rushing towards us at a speed of some 300 kilometres per second. This was far and away the greatest Doppler velocity measured, at the time, for an astronomical object.

The discovery caused amazement, and not a little confusion. Many people were sceptical, until Slipher managed to obtain spectra show-ing what seemed to be large Doppler effects for other spirals, and other astronomers began to confirm his measurements. Some idea of the confusion can be seen from one of the suggestions made by Slip-her at the time — that the high speed of the Andromeda Nebula might explain the "nova" of 1885, if the nebula were a cloud of gas, racing through the Milky Way, that had engulfed a star lying in its path and caused it to explode. The first few spectra obtained by Slipher showed a mixture of blueshifts for some objects (suggesting that those nebu-

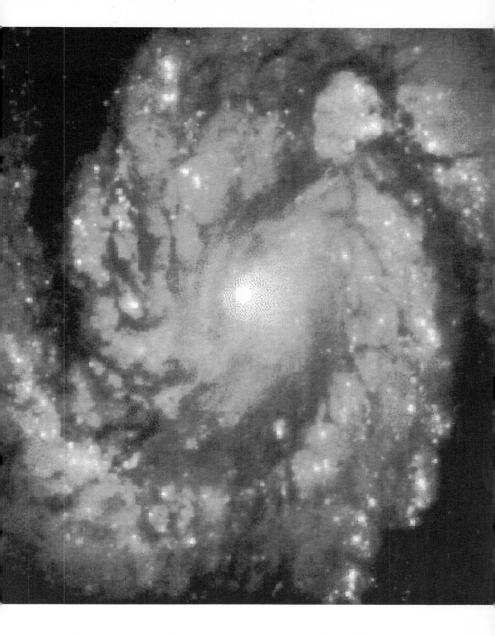

The central region of the spiral galaxy M100, imaged by the Hubble
Space Telescope's Wide Field and Planetary Camera (WFPC2) on
31 December 1993. *(Credit: NASA.)*

Some idea of the power of the
Hubble Space Telescope (pictured in
orbit around the Earth in the NASA
picture above) can be gleaned from
the two images on the facing page.
The main image is a picture of the
barred spiral galaxy NGC 1365
observed from the ground; the detail
shows the small region of this galaxy
inside the white outline, as seen by
the HST. *(Credit: W. Freedman,
HST Key Project Team, and NASA.)*

ABOVE, INSET. A classic example of a kind of
gravitationally lensed image called an Einstein
cross. The four outer bright spots are different
images of the same distant quasar, produced by
light being bent around a spiral galaxy (the
central bright dot) which lies almost exactly on
the line between us and the quasar. *(Credit:
NASA and ESA.)*

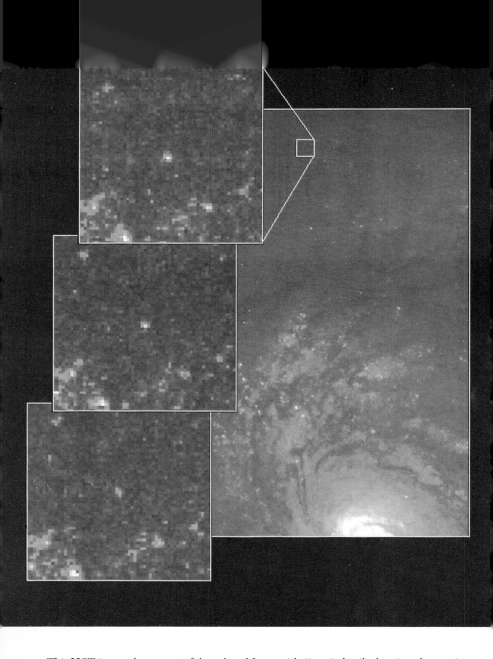

This HST image shows part of the galaxy M100, with (inset) details showing changes in the brightness of a single Cepheid variable star in that galaxy. The three images were obtained in May 1994, and show variations in the brightness of this particular Cepheid over an interval of four weeks. At the time, M100 was the most distant galaxy in which Cepheid variables had been observed. *(Credit: Wendy Freedman and NASA.)*

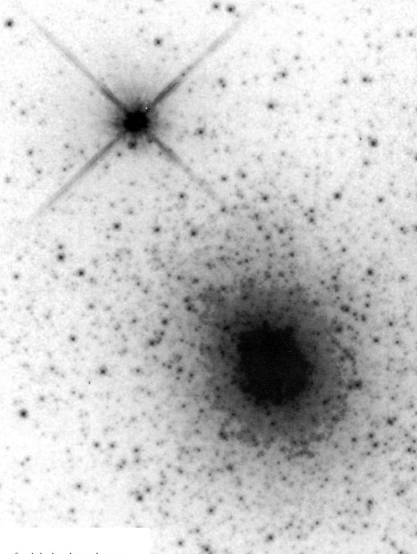

This image of a globular cluster known as G1, photographed by the HST in July 1994, has been printed as a negative, a common trick used by astronomers to highlight details. So stars show up as black dots on a white background. There are more than 300,000 stars in this cluster, which is part of the Andromeda galaxy (M31). *(Credit: M. Rich, K. Mighell, J. Neill, W. Freedman, and NASA.)*

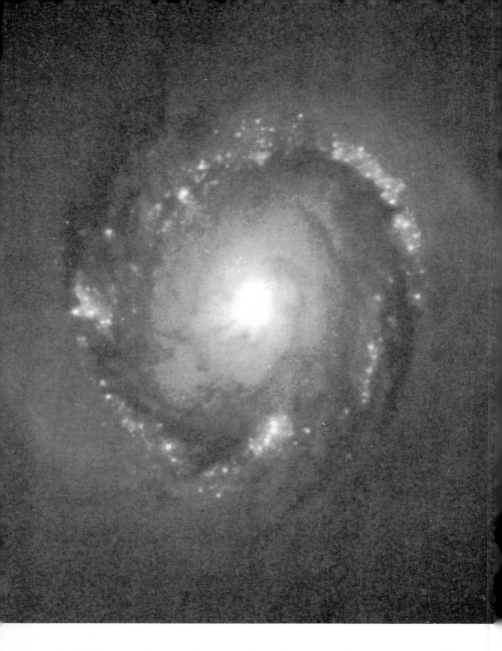

This HST image, released in June 1998, shows a bright ring of young stars around the core of the galaxy NGC 4314. All of the bright stars in the ring are less than five million years old, but nobody knows exactly what triggered the burst of star formation. *(Credit: F. Benedict, A. Howell, I. Jorgensen, D. Chapell, J. Kenney, B. Smith, and NASA.)*

Part of a WFCP2 image of a gravitational lens effect associated with a cluster of galaxies known as 0024 + 1654. The loop-shaped objects on the left of this picture are separate lensed images of a galaxy that lies far behind the bright cluster of galaxies, which is acting as a lens. *(Credit: W. Colley, E. Turner, J. A. Tyson, and NASA.)*

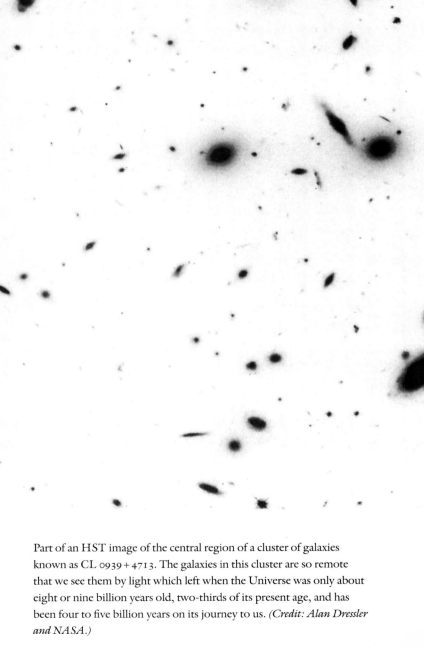

Part of an HST image of the central region of a cluster of galaxies known as CL 0939+4713. The galaxies in this cluster are so remote that we see them by light which left when the Universe was only about eight or nine billion years old, two-thirds of its present age, and has been four to five billion years on its journey to us. *(Credit: Alan Dressler and NASA.)*

lae were moving towards us) and redshifts for other objects (suggesting that they were moving away), which caused more confusion — it seemed at first as if the nebulae were hurtling about through space in random directions, some with even greater velocities than the Andromeda Nebula.

By 1914, Slipher had measured "velocities" for fifteen spirals, including two that had redshifts corresponding to velocities of recession greater than 1,000 kilometres per second. But only two of the fifteen measurements showed a blueshift. The pattern had been established — blueshifts were the exception; redshifts are common. (Before long, blueshifts would be relegated even further in significance; it turned out that the Sun and Solar System are moving through space, in their orbit around the centre of the Milky Way, at a speed of about 250 kilometres per second, almost exactly in the direction of the Andromeda Nebula. All but 50 kilometres per second of the blueshift in the light from the Andromeda Nebula is caused by our motion towards the nebula, not the nebula's motion through space towards us.)

By 1917, Slipher had pushed the number of measured redshifts for spirals up to twenty-three, but there were still only two blueshifts, and by 1925 it was thirty-nine redshifts to two blueshifts. In all this time, other astronomers had measured just ten nebular redshifts in systems that Slipher had not studied first — he had measured four times as many redshifts as everybody else put together. But that was as far as Slipher could go. By then, he was measuring the faintest spectra that could be analysed in this way using the 24-inch refractor and its spectrograph at the Lowell Observatory. It was time for someone else to take up the challenge of getting deeper into the Universe, analysing the light from even fainter nebulae. It was natural that the challenge should be taken up by Hubble and Humason, working with the best telescope in the world. But it isn't always appreciated that the discovery of the famous relationship between redshift and distance did not strike them as a bolt from the blue. Theorists had already suggested that there might be some such link, and Hubble was actively seeking to test these theories in his collaboration with Humason from the mid-1920s onwards.

Remember that Slipher's work on redshifts was essentially complete just before Hubble began measuring the distances to spirals. There was still no consensus on the nature of these objects, but the very large redshifts being measured by Slipher did suggest that they could not be part of the Milky Way proper — whatever the size and composition of these objects, they were simply moving too fast to be held in the gravitational grip of the Milky Way. All kinds of speculations were invoked to account for the phenomenon. Could the nebulae be relatively small clouds of material that were being pushed away from the Milky Way by some process, perhaps even by the pressure of the light of all the stars in the Milky Way (a counsel of despair, but a suggestion put forward seriously at the time)? Could they be some sort of flotsam in the Universe, small objects floating about in space that had been disturbed and pushed aside by the passage of the Milky Way past them, like small dinghies set bobbing about in the ocean by the passage of a great ocean liner? Some astronomers, including Hertzsprung, had no doubt that the high velocities, and the implication that the nebulae are not part of the Milky Way, lent weight to the idea that they were systems like the Milky Way in their own right — but there was always the problem of explaining away van Maanen's rotation measurements. It was against this background of new observations, some of which contradicted one another, that the theories (strictly speaking, one theory in different forms) pointing to the possibility of an expanding Universe developed. But it didn't help that the mathematicians and theoretical physicists who first developed these ideas were not always in close contact with the observers, and didn't realise at first that the equations they played with might have applications to the real world.

It all began in 1916, when Albert Einstein published his general theory of relativity. The general theory is a theory of space, time, and matter. The way it describes space and time can be understood in terms of a four-dimensional geometry, describing spacetime, and we do not need to go into any mathematical details to see the power of this approach as a description of the Universe. Spacetime is a continuous but flexible entity in the general theory, and it can be stretched and distorted by, in particular, the presence of matter. The usual analogy is to replace the four dimensions of spacetime by the more familiar two dimensions of a stretched rubber sheet, like a trampoline. In flat spacetime (the flat trampoline), objects move in straight lines, unless some force acts upon them — a marble rolling across the trampoline, for example, keeps going in a straight line. If a large mass is present, though, spacetime is curved — place a heavy weight on the trampoline and it makes a dent. Now, objects travelling along a line of least resistance in spacetime follow curved paths near the heavy object (a marble rolled around the dent in the trampoline will follow a curved path). The way trajectories are affected by this bending of spacetime is equivalent to the way trajectories are affected by the force of gravity in the older picture of how the Universe works, with the idea of a gravitational force replaced by the idea of curved spacetime. Gravity is entirely due to the distortions in spacetime caused by the presence of matter.

The distortions predicted by the general theory implied, among other things, that light from distant stars, passing near the Sun, should be deflected by a certain amount. Usually, it is impossible to test this, because the light from distant stars is completely lost in the glare of the

Sun in daylight. But in 1919, during a total eclipse of the Sun, careful measurements were made of the apparent positions on the sky of stars along the lines of sight which passed near the obscured Sun. The measurements, made by a team headed by Arthur Eddington, exactly matched the predictions of Einstein's theory, and this was the main reason why that theory was accepted so quickly, at least by mathematical physicists, in spite of its strange predictions (some observers were more reluctant to accept the general theory, as we shall see). It has since been tested many times in many ways, and passed every test with flying colours. We know that the general theory of relativity is a good description of the Universe.

And this is a key point. From the outset, Einstein realised that his theory described the entire Universe — all of space and time, and the way it is affected by all the matter embedded in it. As soon as the general theory was complete, Einstein applied it to the problem of finding a set of equations that would describe the Universe at large. He published the first fruits of those labours in 1917, less than a year after the publication of the general theory itself. But, as that paper makes clear, he had found something very odd about the way the general theory described spacetime at large.

As the trampoline analogy highlights, the general theory allows (or rather, *requires*) spacetime to be stretched and distorted by the presence of matter. Space, as part of spacetime, is itself affected in this way. And Einstein found, in 1917, that the equations of the general theory insisted that space could not be static overall — that it must be either expanding (as if the whole trampoline were being stretched in all directions) or contracting (as if the trampoline were shrinking), but could not be standing still. Such an expansion (or contraction) of space would carry matter along with it, and show up, Einstein thought, in the motions of stars. This was a dilemma, because although Slipher had published his early work on redshifts by then, Einstein was not aware of it, and in any case the received wisdom was that the Milky Way was either the entire Universe, or at least the dominant component of the Universe, and the Milky Way is certainly neither expanding nor contracting. Einstein's resolution of this dilemma was to introduce a new term, denoted by the Greek letter lambda, into his equations in order to cancel out this unwanted motion, and hold his model universe still. As Einstein himself said, "That term is necessary only for the purpose of making possible a quasi-static distribution of matter, as required by the fact of the small velocities of the stars." So, thanks to the lambda term (sometimes known as the cosmological constant) the cosmological model he published in 1917 actually did describe mathematically a space which stayed the same size, without expanding or contracting.

Almost immediately, though, the Dutch astronomer Willem de Sitter found another solution to the equations of the general theory, which also described a complete universe — but a different kind of universe from the one Einstein had described. De Sitter found a mathematical model which described a completely empty universe with no matter in it at all. The spacetime of this empty universe stayed still all by itself without any need for a lambda term. But if any matter at all were added to the model universe (even a few grains of sand) it began to expand violently. In fact, it expanded exponentially fast. In this context, exponential means that you multiply the relevant numbers up by the appropriate "power," or exponent, so that 2^2 becomes 2×2, 4^2 becomes 4×4, and so on. If the exponent involved really is "2," for example, this means that an object twice as far away from you is moving away not twice as fast, but four times (2^2 times) as fast as a nearer object; while an object four times further away recedes sixteen times (4^2 times) as fast, and so on.

As early as 1917, this de Sitter model did predict a redshift in light from distant objects — but in the model, this redshift is interpreted as caused by a change in the properties of spacetime from one part of the universe to another which makes distant clocks run more slowly, and distant atoms radiate light in a more leisurely fashion, regardless of their motion. But the model did predict that more distant objects should show larger redshifts, and this was the first time the idea of a redshift–distance relation was introduced into astronomy. In 1917, de Sitter (in war-torn Europe) had only heard of the first three "Doppler" measurements made by Slipher, and one of those was the huge blueshift of the Andromeda Nebula, so it was not obvious that his model had any bearing on the real world. When Slipher's first twenty-five measurements (most of them redshifts) were available, in the early 1920s, even de Sitter himself did not claim this as evidence in favour of the model, because nobody knew the distances to the spirals being studied by Slipher, and without distances how can you measure a redshift–distance relation?

Even in the first half of the 1920s, though, several attempts were made to link the observations of redshifts in spirals with the idea of what we would now call the expanding Universe. Most of these efforts came from mathematical physicists, who were interested in working out solutions to the equations of the general theory of relativity. The attitude of many observers is indicated by George Ellery Hale's comment, in 1920, that "the complications of the theory of relativity are altogether too much for my comprehension. If I were a good mathematician I might have some hope of forming a feeble conception of the principle, but as it is I fear it will always remain beyond my grasp." The mathematicians, though, seem to have had an almost touching faith in the limited amount of observational data. Their investigations were based on a minimal amount of observations — the

few redshifts measured by Slipher, and vague distance estimates rang-
ing from observations of novae to the reasonable guesses that more
distant spirals might be fainter than nearby ones, and look smaller
on the sky. These very rough-and-ready methods enabled the Polish
mathematician Ludvik Silberstein (then based in England) to claim
evidence in favour of a redshift–distance relation in 1924, and this
encouraged the Swedish astronomer Knut Lundmark (a long-time
advocate, as we have seen, of the idea that spirals are galaxies in their
own right) to carry out a more honest analysis (Silberstein tended to
leave out any observations that didn't match his model) which concluded, in 1925, that there might be a redshift–distance relation, but
"not a very definite one."

This is probably the most honest and accurate appraisal of the
situation in the mid-1920s, around the time that Hubble established
the distance to the Andromeda Nebula and a few other spirals. There
were other stabs at the redshift–distance puzzle in the mid-1920s, but
none led to any firm and convincing conclusions. One of the leaders
in the field was Carl Wirtz, a German astronomer who found, in 1924,
what Hubble himself later called (in his book *The Realm of the Nebulae*) "a plausible correlation . . . velocities tended to increase as diameters diminished." But as Hubble went on to say, "The results, however, were suggestive rather than definitive." We don't need to go into
all the details, but I do want to make it clear that the idea of a redshift–
distance relation was in the air before Hubble and Humason began
their epic work investigating this relation, and that the myth that they
discovered the relation between redshifts and distances "out of the
blue" is just that — a myth. The importance of their work, as we shall
see, is that it removed all the vagueness, the qualifying terms like
"plausible" and "tendency," and put the whole idea on a secure, quali-
tative footing. But it is true that they, along with almost every other

astronomer at the time, were entirely ignorant of two pieces of work which had been published in the 1920s, and which, if they had been more widely known, would have put the discoveries of Hubble and Humason more clearly in context from the outset.

It is one of the greatest mysteries of the history of science in the twentieth century why the cosmological work of the Russian mathematician Alexander Friedman made almost no impact on astronomy when it was published in 1922. Friedman's key paper was published in one of the most widely read and prestigious scientific journals (the German *Zeitschrift für Physik*), and even drew a response from Einstein, who thought he had spotted a mistake in the calculations but later admitted that he was wrong and Friedman was right. But the trouble was that Friedman presented his solutions to the equations of the general theory of relativity as mathematical curios, rather than suggesting that they might have any bearing on the real physical Universe. As for Einstein, if anything he was annoyed by the paper, because it seemed to undermine the simplicity of his own version of relativistic cosmology.

What Einstein had tried to do in his own work in 1917 was to find a *unique* cosmological model, the only one allowed by the general theory, and then to compare this with the real world. As de Sitter's work had already hinted, this was not to be. There was more than one way to interpret the equations of the general theory. This is now seen as a good thing, an example of the richness of the general theory; to Einstein in the early 1920s, it seemed a bad thing, because he had hoped that his theory would uniquely match up to the real world, telling us that only a Universe like our own could exist. Good or bad, though, it was Friedman who first spelled out the way in which Einstein's equations give rise to a whole family of solutions — a family of cosmological models, or different universes — which describe different possible ways in which spacetime can behave. In some of the models, the universe expands forever as space stretches. In other variations on the theme, it expands to a certain size, then collapses, shrinking down towards a point, where it may "bounce" into another cycle of expansion and collapse. And, by including a cosmological constant just as Einstein had done, Friedman could produce models which stayed the same size for eternity.

In a sense, Friedman gave astronomers too much, too soon. He did not predict a unique expanding Universe, which could then be compared to observations of the real Universe, but offered a wealth of choices, which seemed to be all things to all astronomers. You could, it must have seemed to those few astronomers who did notice the work, make any observations fit one of Friedman's models, so there didn't seem much point in the exercise. And, besides, Friedman died in 1925, and wasn't around to champion his ideas. But that wasn't the end of the story.

In the second half of the 1920s all of Friedman's results were independently rediscovered by the Belgian astronomer Georges Lemaître (in total ignorance of Friedman's work), who travelled widely in Europe and the United States (Friedman had never been able to leave the Soviet Union) and who, unlike Friedman, was seriously interested in relating the mathematical equations to the real Universe. In some ways, then, it is much more surprising that his results went equally unsung for years. But Lemaître chose to publish his greatest work in an obscure Belgian journal, and for a long time he did not promote the ideas described in his paper on his visits to people like Eddington, Slipher, and even Hubble.

Lemaître was another astronomer whose life and career had been disrupted by World War One, although much more violently than Hubble's had been. Born in 1894, Lemaître had been studying engineering at the Catholic University in Louvain, but when the German army invaded Belgium in August 1914 he volunteered for the army and was involved in front-line combat, being awarded the Croix de Guerre avec palmes. In 1919, he took up his studies again, switching to mathematics and physics and obtaining a doctorate (about the equivalent of a modern master's degree) in 1920; he then joined a seminary and became an ordained priest in 1923. Lemaître never practised as a parish priest, but rose in the clerical establishment because of his scientific interests, becoming a member of the Pontifical Academy of Sciences in 1936, and serving as its President from 1960 until his death, in 1966. What matters for our own story, though, is that immediately after leaving the seminary Lemaître spent a year in Cambridge, working with Arthur Eddington, then visited the United States, where he worked with Shapley at Harvard (working for a Ph.D. in astronomy, a more prestigious qualification than his existing one) and visited both Slipher and Hubble at their own observatories — expressly because he was interested in the application of relativity theory to a description of the real Universe, and appreciated the importance of both the redshift measurements and the distance measurements then being carried out. He was also present at the AAS / AAAS joint meeting where Hubble's paper announcing the discovery of Cepheids in the Andromeda Nebula was read.

Lemaître returned to Belgium, transformed by his investigations during his travels from a mathematical physicist into a cosmologist with a thorough understanding of the latest developments in observational astronomy, and was appointed Professor of Astronomy in Louvain in 1927, the year that he published his now classic paper on the expanding Universe.

Although Lemaître essentially reproduced all of the discoveries made by Friedman (with some differences that need not concern us here), the key difference between the two researchers lay in their approach. Friedman was a mathematician playing with the equations of the general theory; Lemaître was (by 1927) an astronomer trying to find a description of the real Universe. There is no more telling indication of the difference between them than the fact that the word *galaxy* (or *nebula*) does not appear anywhere in Friedman's paper, but it does in Lemaître's. Lemaître spelled out that the expansion of the Universe, with space stretching and carrying galaxies further apart as time passed, would cause a redshift in the light from distant galaxies (but there was no Big Bang in the 1927 version of Lemaître's work; his preferred model started out as a stationary universe, like Einstein's model, and then, after an indefinite amount of time, started expanding).

In his 1927 paper, Lemaître even quoted a figure for the relationship between redshift and distance, a constant of proportionality amounting to 625 kilometres per second per Megaparsec (in other words, a galaxy 1 Mpc away will be receding at 625 km/sec; a galaxy 2 Mpc away will be receding at 1,250 km/sec; and so on). Lemaître did not say where this number came from — but he had not long returned from the United States, where he had visited Hubble, who was already working on the redshift–distance relation, and this figure is quite close to the one that Hubble would publish two years later. So close, indeed, that the cosmologist Jim Peebles, writing in 1971 in his book *Physical Cosmology,* said that "there must have been communication of some sort between the two."

There was certainly some communication between Lemaître and Einstein, since Lemaître later (in a memoir published in 1958) said that he had shown his paper to Einstein in 1927, and Einstein had agreed that it was sound, but pointed out that Friedman had already reached much the same conclusions. Einstein also vigorously opposed the idea that the Universe might be physically expanding, preferring the kind of interpretation used in de Sitter's model, with redshifts being caused by differences in the structure of spacetime at different distances. It may have been this discussion with Einstein which dissuaded Lemaître from banging the drum about his work—in any case, he had visited the United States again in the spring of 1927, to complete the formalities for his Ph.D. from Harvard, and then took up his post as Professor in Louvain, so he probably had plenty of other things to worry about at the time.

The real oddity, though, is that if there was enough communication from Hubble to Lemaître for the first public appearance of what is now called Hubble's Constant to have been in the 1927 paper by Lemaître, why was there so little communication the other way that Hubble didn't mention Lemaître's work a couple of years later? Didn't Lemaître even send a copy of his paper to Hubble? We shall never know. But we do know that it was a paper by Hubble, published in 1929, that finally made the astronomical community sit up and take notice of the idea that the Universe is expanding, with the implication that it was born at a definite moment in time, and has an age which we can calculate. And, as the link with Lemaître shows, Hubble had been worrying at the puzzle for several years before 1929.

Hubble actually mentioned the de Sitter model of the Universe in one of his first papers on distances to nebulae, published in 1926. But the distances themselves continued to occupy his attention long after the breakthrough identification of Cepheids in the Andromeda Nebula. Cepheids themselves are only just bright enough to give distances to the Andromeda Nebula itself and a very few other nearby galaxies, and in order to probe deeper into the Universe Hubble had to use a variety of secondary techniques, all time-consuming and none as reliable as the Cepheid technique. Novae, for example, can be seen a little further out into the Universe than Cepheids can, and since the distance to the Andromeda Nebula had been measured from Cepheids, the average brightness of all the novae seen in the Andromeda Nebula could be used as a benchmark for calculating the absolute brightnesses of more distant novae and thereby inferring, from their apparent brightnesses, the distances to their host galaxies. Hubble also used the very brightest individual stars in galaxies (much brighter than Cepheids) as distance indicators, guessing that there must be some upper limit on how bright a star could be without exploding, so that the brightest stars in the Andromeda Nebula must be intrinsically the same brightness as the brightest stars in more distant galaxies, so again the apparent brightnesses of those more distant stars could be used as distance indicators. As I have mentioned, a rough-and-ready distance indicator could be based on the apparent size of a galaxy on the sky (the more distant a galaxy is, the smaller it looks), and even the brightness of a whole galaxy might be compared with the brightness of the Andromeda Nebula or some other nearby galaxy, to get an indication of distance (if all galaxies had the same brightness, which unfortunately they do not, a galaxy one hundredth as bright as a chosen nearby galaxy would be ten times further away).

With the obvious uncertainties involved in all these estimates, the best way to get a reasonable guide to the distance to any individual galaxy (or a cluster of individual galaxies moving together through space) was to apply as many of the different techniques as possible to each galaxy (or each cluster). It all took time, but gradually Hubble began to build up a catalogue of distances to nebulae. Only then was he really in a position to turn his attention to redshifts, and try to find a relationship between redshift and distance. And what was at the back of his mind when he began this work was the idea that if redshift really is proportional to distance, then all that he would have to do to measure distances in future would be to measure the redshifts, and multiply by the appropriate constant of proportionality.

Just at the time that Hubble had identified the nature of the Andromeda Nebula and other spirals as galaxies beyond the Milky Way, in 1926, Slipher was coming to the end of his studies of redshifts, because the equipment he had available, based on the 24-inch refractor, had been pushed to the limit of what it could observe. If Hubble wanted to search for a relationship between redshift and distance, the first thing he would have to do would be to find distances to as many as possible of the nebulae whose redshifts had been measured by Slipher. But in order to probe deeper into the Universe, as he realised in 1926, he would need redshifts for fainter objects, which could best be obtained by the 100-inch. Hubble himself was deeply involved with the continuing programme of distance measurements, and the 100-inch had never been used for redshift work, involving spectroscopic photographs of very faint objects, before. He needed someone to undertake the taxing task of adapting the telescope to this new work, and then making the measurements themselves.

Humason was the obvious choice, not just because he was a superb observer, but also because of the clear difference in status. Although Hubble knew he had to have help with his latest project, he didn't want a collaborator of equal status as an astronomer to himself; he wanted an assistant, so that as much as possible (preferably all) of the glory associated with the work would be his.

Humason took up the challenge, and in order to test the possibilities he chose for his first attempt at a redshift measurement a nebula which was too faint for its light to have been analysed in this way by Slipher at the Lowell Observatory. After two nights patiently keeping the great telescope tracing the faint nebula, he had a spectrum good enough to show (under a magnifying lens) spectroscopic lines associated with the presence of calcium atoms in the nebula. The lines were shifted towards the red end of the spectrum, by an amount corresponding to a Doppler velocity of some 3,000 kilometres per second, more than twice as large as any redshift measured by Slipher.

The trial run had been a success, but it had also shown Humason how physically demanding it would be to obtain more spectra from faint nebulae. The prospect of spending night after night freezing in his seat at the guidance controls of the telescope, all for the benefit of someone else's research project, and all to confirm (at least at first) what Slipher had already discovered, did not appeal to him, and he said so in no uncertain terms. He was persuaded to carry on with the task partly by some flattering comments from Hale (who had retired as Director of the Mount Wilson Observatory, on health grounds, but still kept in close touch) and by the promise of a new spectrograph, much more sensitive than the old one, which would enable spectra of even faint nebulae to be obtained in a single night. Humason agreed to carry on. Of course, in the long run the new spectrograph didn't really ease his burden. If a faint nebula could now be photographed spectroscopically in a single night, then a *very* faint nebula could be photographed in two or three nights of observation. Astronomers are always pushing their equipment (and in those days, themselves) to the limit. Before long, Humason was hooked on the project, working harder than ever to obtain redshifts for fainter and fainter objects.

But he took things step by step. Showing exemplary caution and patience (he must have been a really good mule driver), in spite of his initial success Humason spent many months bedding the new equipment in, and honing his own skill at the new technique, by remeasuring the redshifts of all forty-five nebulae analysed by Slipher. He found the same values of the redshifts that Slipher had found, important confirmation that the results meant something (remember the puzzle over van Maanen's rotation measurements), and that the combination of the 100-inch, the new spectrograph, and Humason himself was ready to take the leap out to higher redshifts.

Meanwhile, Hubble had been making distance measurements (using the variety of techniques I have outlined) for many of the same nebulae, and had a pretty good idea that the two sets of data showed a linear relationship between redshift and distance — that redshift is proportional to distance, so that if one galaxy has twice as big a redshift as another, it is twice as far away. Indeed, he must have had some idea of this already in 1926, as we have seen from the evidence of Lemaître's 1927 paper; but he was extremely cautious about putting this conclusion down in print, and was only pushed into doing so when it looked as if someone else was on the same trail.

The someone else was Lundmark, who at the end of 1928 made a formal request to the then Director of the Mount Wilson Observatory, Walter Adams, to visit the mountain once again, for the express purpose of measuring the redshifts of faint nebulae. He even asked if Milton Humason might be available to help him in this work. Lundmark was politely rebuffed, and Hubble took the hint, publishing his first short paper on the redshift–distance relationship early in 1929. In that paper (just six pages long, and titled "A Relation Between Distance and Radial Velocity Among Extra-Galactic Nebulae") Hubble claimed to have accurate distance measurements to just twenty-four of the forty-six nebulae for which redshifts were widely known at the time, and less accurate distances to the other twenty-two. When these measurements were plotted as points on a graph, with distance along the horizontal axis and velocity up the vertical axis, they were scattered rather widely, but with a tendency for higher velocities to be associated with higher redshifts. Hubble drew a straight line through these scattered points, with a slope which set the constant of proportionality in the redshift–distance relation as about 525 kilometres per second per Megaparsec (about 20 percent less than the value quoted by Lemaître in 1927).

On the evidence of the 1929 paper alone, it is hard to justify choosing this particular slope for the straight line (to be honest, it is hard to justify drawing a straight line at all); but Hubble already knew of at least one galaxy with a much higher redshift and correspondingly greater distance, and it is certain that he chose this particular straight line to make his published results in that 1929 paper line up with the unpublished data for larger redshifts that he was still working on. Why was he so cautious about revealing the new results that were now coming in from a comparison of his own distance work and Humason's redshifts? Because he wanted to finish the job before publishing a full paper. If other astronomers (such as Lundmark) got wind of just how successful Humason was being in his measurements of very high redshifts, they might get in on the act, and steal some of the thunder from the Mount Wilson team. Sharing the glory with Humason, clearly his junior, might just be acceptable; sharing the glory with someone from a different observatory was not.

Even so, Hubble's claim of a linear redshift–distance relationship was quickly accepted by the astronomical community, and became known as Hubble's Law. After all, as we have seen, the idea of some sort of relationship between redshift and distance was very much in the wind, and people were primed to believe it (not least because a linear relationship is the simplest kind, and the easiest to work with). The snag was, that the kind of redshift–distance relation found by Hubble (and the as-yet-unsung Humason) did not match up with either the Einstein model of the Universe or the de Sitter model. Eddington commented on this difficulty for the theorists at a meeting of the Royal Astronomical Society, in London, in January 1930 (he had completely forgotten, if he had ever read, Lemaître's 1927 paper). When Lemaître read these comments in the published account of the meeting, he wrote to Eddington, enclosing another copy of the paper and pointing out that the kind of redshift–distance relation found by Hubble could indeed arise naturally in the context of the general theory of relativity. This time, Eddington not only read the paper, but promptly wrote to *Nature,* the leading scientific journal of the time, drawing attention to Lemaître's work; he also sent a copy on to de Sitter, who enthusiastically endorsed it. Almost everyone agreed that Lemaître had the explanation for the redshift–distance relation discovered by Hubble, and that the Universe as a whole must be physically expanding, getting bigger as time passes.

At the beginning of 1931, Hubble and Humason together published a paper, "The Velocity–Distance Relation Among Extra-Galactic Nebulae," which at last revealed most of the data which Hubble had been hugging to himself for the past couple of years. With another fifty redshifts, they more than doubled the number in Hubble's 1929 paper, and pushed the record out to a cluster of galaxies with a redshift corresponding to a velocity of recession of just under 20,000 kilometres per second, at a distance estimated at the time as a little more than 100 million light-years. When the data were plotted as a graph, the straight line, with almost the same slope as in the 1929 paper, was still there; but the scatter in the points along the line was much smaller, and the choice of the slope for the straight line looked much more plausible.

But what did it all mean? If the galaxies are moving apart today, the implication is that they were closer together in the past. Go back far enough into the past, and they must have been on top of one another — there must have been, in some sense, a beginning to the expanding Universe. If the expansion has been going on at the same rate all the time, it is easy to calculate, from the constant of proportionality in the redshift–distance relation, how long it has been since all the galaxies in the visible Universe were squashed together in one lump. Using a value of about 525 kilometres per second per Megaparsec for the constant (which became known as Hubble's Constant, and is now denoted by the letter H, although Hubble himself used K), this "age of the Universe" comes out as about two billion years.

By the 1930s, as we have seen, radioactive dating techniques had already established that the Earth and the Solar System must be much older than this. Either those measurements were wrong (unlikely), or the Universe had not always been expanding at the same rate (quite plausible; remember Lemaître's idea of a universe that "hovers" indefinitely before expanding), or there was something wrong with Hubble's interpretation of Humason's redshift measurements. But the important point is that such questions were being asked in the early 1930s, not that there were yet any definite answers. Before the discovery of the redshift–distance relation, Hubble's Law, nobody seriously considered the possibility that the Universe itself had been born at a definite moment in time and had a measurable age. After the 1931 paper by Hubble and Humason, the idea of measuring the age of the Universe became part of the scientific investigation of the world.

Hubble himself was always cautious about the physical meaning of these discoveries. Although it was natural to interpret the redshift as a Doppler effect, caused by the motion of galaxies through space, he avoided expressing any opinion on the matter. In his classic book *The Realm of the Nebulae,* published in 1936, he wrote: "Red-shifts may be expressed on a scale of velocities as a matter of convenience. They behave as velocity-shifts behave and they are very simply represented on the same familiar scale, regardless of the ultimate interpretation.

The term 'apparent velocity' may be used in carefully considered statements, and the adjective always implied where it is omitted in general usage."

He was right, and the adjective is indeed always implied in the discussion of redshifts in this book. The redshifts are associated with *apparent* velocities, not real movement of galaxies through space, as we have seen. They are *not* caused by the Doppler effect, but are, in the language of the general theory of relativity, caused by a time variation of the metric scale factor in the Lemaître equation. In more ordinary language, the interpretations of the redshift effect that were already being made in the 1930s, when Hubble wrote the words quoted above, did not involve velocities in the everyday sense, but the stretching of space itself. And those interpretations pointed, ever more clearly, to the idea that there is indeed a definite age of the Universe.

6

Revisionist Cosmology
Extending the Age of the Universe

It was the combination of the announcement of the redshift–distance relation by Hubble and Humason, and the rediscovery of the expanding Universe models of Lemaître that made the study of the expanding Universe part of mainstream science, from the early 1930s onward. The combination of theory and observation (equivalent to experiments) showed scientists that they were dealing with a real phenomenon. Without the theory, nobody would have known what to make of the redshift measurements (indeed, even with the theory a few people tried to explain the redshifts in other ways, but without success). Without the observations, nobody would have known what the theoretical models were all about. Put the two together, and you had the basis of a description of the real Universe. The combination was so sensational and exciting that the resulting debate even made the pages of *The Times,* where, in a series of contributions throughout May 1932, the astronomer and science populariser James Jeans attempted to explain to baffled readers the notion of curved and expanding spacetime.

In the 1930s and 1940s, the main cosmological developments were in interpreting and explaining the ideas which developed from these two great discoveries of the 1920s. Both these lines of attack look equally important with hindsight; but one of them was largely ignored at first. This may have been because even astronomers and cosmologists were still uncomfortable with the new idea that the Universe may have had a definite beginning, a finite time ago (and part of the reason for this discomfort must have been that if you did plug Hubble's own numbers for the redshift–distance relation into the simplest cosmological models, the resulting "age of the Universe" came out as less than the age of the Earth). This reluctance to accept what the combination of observation and theory was telling them is highlighted in a talk Eddington gave to the British Mathematical Association, in January 1931. His address, as President of the Association, mainly dealt with the idea of entropy, which says that the amount of disorder in the Universe is always increasing ("things wear out"), and that it faces a "heat death" in which everything has been smoothed out into a uniform state at a uniform temperature. But if you imagined reversing the process, and winding the history of the Universe backwards, could there also have been a "heat birth," a state of perfect order in which the Universe was created? "The notion of a beginning," said Eddington, "is repugnant to me."

That comment encouraged Lemaître to develop his cosmological ideas further, and to point out that science did not depend on personal taste, but on the results of experiments and observation. The idea of a beginning might be repugnant, but if that is what is implied by the observations, it has to be taken seriously. Lemaître took up the challenge of providing a scientific description of what the beginning might have been like (a model) by drawing on the new ideas that had emerged from quantum physics in the 1920s. In a "letter" to the scientific journal *Nature* (*Nature* calls its scientific papers "letters"), specifically responding to Eddington's comment, Lemaître said that the beginning of the world "is far enough from the present order of nature to be not at all repugnant," and went on to propose that: "We could conceive the beginning of the universe in the form of a unique atom, the atomic weight of which is the total mass of the universe. This highly unstable atom would divide in smaller and smaller atoms by a kind of super-radioactive process."

In other words, the Universe could have got to be the way it is by the repeated fission of the super atom (like a huge atomic bomb) — or rather, by the fission of a super *nucleus*. What Lemaître was really talking about was a single atomic nucleus containing all of the mass of the visible Universe, which exploded at the beginning of time. Such an object would have been only about thirty times bigger across than our Sun, but would have the density of the nucleus of an atom — which gives you some idea of just how much empty space there is between the stars and galaxies (indeed, how much empty space there is in an atom of ordinary matter, where the tiny, central nucleus containing essentially all the mass is surrounded by a cloud of electrons as remote from the nucleus as the dome of St. Peter's in Rome is from a speck of dust on the floor of the church).

This idea became known as the primeval atom, and although it was based on speculation and metaphor rather than on hard and fast equations, Lemaître soon linked it to his cosmological models, using a cosmological constant (a lambda term). In the particular model he favoured, the Universe began with the explosion of the primal atom, expanded for a while, then hovered for much longer in a more or less stationary state (removing any problem of having the age of the

Universe less than the age of the Earth) before expanding once again. On that picture, we would be living in the second phase of expansion of the Universe. Lemaître became something of a celebrity as a result of his elaboration of these ideas both in scientific papers and several popular contributions, culminating in his book *L'Hypothèse de l'Atome Primitif (Hypothesis of the Primal Atom)*, published in 1946. This acclaim looks amply justified today, because whatever its faults, Lemaître's favoured cosmological model was the first time anyone made a serious scientific attempt to explain what went on "in the beginning," and tried to reconcile calculations of the age of the Universe based on redshift data with estimates of the age of the Earth.

But the primeval atom idea was largely ignored by Lemaître's fellow scientists in the 1930s, and although it can now be seen as a forerunner of the Big Bang idea, Big Bang cosmology really only took off in the 1940s, as a result of the work of George Gamow and his colleagues. I have told the story of how things developed following Gamow's work in my book *In Search of the Big Bang*, and I will say no more about it here, since in this book I am focusing on the age of the Universe debate. But perhaps it is worth pointing out that there was a downside to Lemaître's popularisation of the primal atom idea. Unfortunately, the image conjured up is of the primal atom (or nucleus) sitting in empty space, and then exploding out into the void. This is wrong; as Lemaître himself was well aware, whatever did happen in the beginning involved the birth of space and time, as well as of matter and energy. There was no "outside" for the primal nucleus to explode into, and when the Universe did indeed have the density of the nucleus of an atom, the entire visible Universe, all of the space we can now "see," occupied a volume only thirty times larger than our Sun. This was the kind of idea that the *Times* readers found so difficult to grasp, even with the aid of Jeans' explanations.

Lemaître's version of an expanding cosmological model was taken seriously by his scientific colleagues in the 1930s (more seriously than the primal atom idea), but as just that — one solution among many to the equations of the general theory of relativity, not as being particularly significant. There was, indeed, a lot of interest in the various solutions to Einstein's equations among the mathematicians in the wake of the discovery of the expanding Universe. In particular, the American mathematician Howard Robertson and his British colleague Arthur Walker developed a whole family of mathematical models which are particularly convenient to work with (in terms of manipulating and interpreting the equations), and these Robertson–Walker models were not only regarded as important in the 1930s, but are still (unlike Lemaître's model) discussed and used today. But there was one specific model of the Universe which proved particularly useful in the context of the age of the Universe debate, and which emerged as early as 1932, hot on the heels of the primeval atom idea. This was the second leg of what became the Big Bang theory, and it is the model most relevant to the work described later in this book. It was developed by two scientists who had already each made individual contributions to the cosmological debate — Albert Einstein and Willem de Sitter.

Einstein, in fact, had endorsed the expanding Universe idea, and abandoned the cosmological constant, as early as April 1931, when he visited the Mount Wilson Observatory during a prolonged trip to the United States, and learnt about the redshift work first-hand. Back in Europe, the following year (1932) he worked with de Sitter on a new model of the Universe, also based on a solution to the equations of the general theory of relativity. This became known as the Einstein–de Sitter model, but it is important to appreciate that it is completely different from either Einstein's original stationary cosmological model or de Sitter's original exponentially expanding cosmological model. The key feature of the Einstein–de Sitter model is that it expands at very nearly a steady rate, with redshift proportional to distance, just like the expansion seen in the real Universe. And from the outset, unlike many of the mathematically inspired models of the 1930s, the whole *raison d'être* of the Einstein–de Sitter model was based on observation.

The two theorists started out from the fact that the only information that could be determined about the Universe at large directly from the observations available at that time were its rate of expansion, and its density. The expansion rate we already know about — it is the number now known as Hubble's Constant in the redshift–distance relation. The density of the Universe could (it was hoped in 1932) be determined by measuring (or estimating) the number of galaxies in a particular volume of space, allowing for the number of stars in a typical galaxy, and working out what the average density of matter would be if all the material in all the stars and galaxies were spread uniformly through space.

One of the key things the mathematicians were interested in was the nature of the curvature of space allowed (or, indeed, required) by the general theory of relativity. There were two obvious possibilities. Either space (or spacetime) itself could have positive curvature (the equivalent in three — or four — dimensions of the way that the surface of a sphere curves around upon itself), or it could have negative curvature (a slightly harder concept to picture, but rather like the shape of a saddle, which curves outwards and away, in principle forever). Each of these alternatives gives rise to a whole family of different cosmological models which (for example) expand at different rates. But in each case all the members of one of the families share a common feature. The first family of possibilities would correspond to a set of universes which each expanded for a time, but in which gravity eventually overcame the expansion and made the universe collapse once again, perhaps with a "bounce" at very high density. The second family of possibilities would correspond to a set of universes which each expanded forever, even though the expansion would get slower and slower as time passed. For obvious reasons, the first kind of universe is said to be "closed," and the second kind is said to be "open."

But there was one more possibility, a special case, a unique solution to Einstein's equations. Einstein had originally been trying to find a unique solution, and now, working with de Sitter, he had one. If the cosmological constant could be set to zero, they wondered whether the curvature could be set to zero as well, simplifying the equations still further. This corresponds to a so-called flat space (or spacetime), mathematically equivalent to the flat surface of a sheet of paper lying on a desk. The appeal of the model (particularly to Einstein) was (and is) that if the expansion rate is known (from the measured redshift–distance relation), then there is only one possible Einstein–de Sitter model — it is unique. This corresponds to the fact that there are many ways in which a sheet of paper could be bent and crumpled up, but only one way in which it can lie smooth and flat on the desk. In terms of expansion, it corresponds to a universe that expands at an ever decreasing rate, until in the far, far future it is left hovering in a delicately balanced state, neither expanding nor contracting. This is the Einstein–de Sitter model. And this is only possible if the density of the universe has a certain critical value, so that the strength of gravity (trying to halt the expansion) exactly balances the actual rate at which the universe is expanding. In the Einstein–de Sitter model, once the expansion rate was known the density could be calculated, and the number they came up with (using the original value of Hubble's Constant, just over 500 km/sec per Megaparsec) was a density of just 4×10^{-28} grams for every cubic centimetre of space. This is, indeed, close to the kind of rough number you get if you count stars and galaxies and estimate the actual density of our Universe. Modern estimates of the critical density are (because modern estimates of H are smaller, for reasons which will become clear later), slightly lower, between 10^{-29} and 2×10^{-29} grams per cubic centimetre; that corresponds to about one atom of hydrogen in every million cubic centimetres, if all the atoms were distributed evenly through space.

In many ways, the Einstein–de Sitter model represents the simplest solution to the cosmological equations of the general theory of relativity (at least, the simplest one that looks anything like the real Universe), and for this reason it became the standard model against which ideas could be tested. This does not mean that it is accepted as the definitive description of the actual Universe, but rather that it is a benchmark against which the way the real Universe operates can be compared. For example, when observations of the real Universe began to suggest, in the 1950s and beyond, that there could not be enough visible matter in all the stars and galaxies to make the real Universe flat in this sense, the usual way to describe this was in terms of the critical density that appears in the Einstein–de Sitter model. Instead of talking in terms of grams per cubic centimetre, or atoms per million cubic centimetres, cosmologists talk of universal densities as being 0.1, or 0.3, or whatever the appropriate number is, of the critical density.

But one thing Einstein and de Sitter carefully avoided discussing in their paper in 1932 was the implication from their model of a definite origin of the Universe, a finite time ago. Of course, they were aware of the age problem — de Sitter, in particular, was actively involved in the debate, and argued in other papers that the Universe must be older than the few billion years implied by Hubble's original interpretation of the redshift–distance relation. No hint of this appeared in their paper. To later generations of cosmologists, though, one of the most appealing features of the Einstein–de Sitter model is that it gives a very simple relationship for calculating the age of the Universe. In this model, the age of the Universe is just two-thirds of the age calculated by assuming that it has been expanding at the same rate we see today ever since the beginning (an age known as the Hubble time, which, remember, itself goes as one divided by the value of Hubble's Constant). This is because the Universe was expanding faster when it was younger, so it has taken less time to reach its present state than you would calculate from present-day observations of the redshift– distance relation (and it has slowed down, of course, simply because of the effect of gravity). Putting it another way, Hubble's "Constant" was bigger in the past, which is why it is sometimes called Hubble's Parameter, with the term Hubble's Constant meaning the value of the Hubble Parameter at the present day.

So there is a fairly narrow range of possible ages for different cosmological models, provided the cosmological constant is zero. The minimum age is given by the Einstein–de Sitter model, and this is just two-thirds of the maximum age, which we get by assuming the expansion has always gone on at the same rate (even shorter ages could result if there were much more matter in the Universe, making it closed; but there is no evidence for this at all). This is a beautiful example of the simplicity and power of the Einstein–de Sitter model; but it is hardly any wonder that the two cosmologists were too embarrassed to discuss it in their 1932 paper, since it meant that if the real Universe were indeed flat and described by the Einstein–de Sitter model, and if Hubble's determination of the redshift–distance relation was correct, then the age of the Universe was just 1.2 billion years, scarcely a third of the then well-established minimum age of the Earth.

One resolution of the puzzle might have been to suggest that there was something wrong with Hubble's determination of the redshift–distance relation, but this was not a widely touted idea at the time. There was, though, another piece of evidence that there was something odd about the scale of Hubble's universe, and at least one astronomer was secure enough, and independent enough, to point this out. By measuring the distances to other galaxies, Hubble and his successors were also obtaining information about their sizes. A small galaxy very close to the Milky Way Galaxy in the Universe will look bigger on the sky than a large galaxy a long way away — just as a child standing next to you looks bigger than an adult on the other side of a football field; a very simple perspective effect familiar from everyday life. The distances found by Hubble himself implied that the other spiral galaxies (or nebulae) he had studied, beyond the Milky Way, were indeed rather small, and rather close to us. This is reflected in the high value he found for the Hubble Constant. If the constant is as high as 500 km/sec per Megaparsec, then a measured redshift of 500 km/sec would indeed correspond to a distance of 1 Mpc. But if the value of the constant were only 100 km/sec per Megaparsec (say), then a measured redshift of 500 km/sec would correspond to a distance of 5 Mpc, and a galaxy five times further away from us would have to be correspondingly bigger in terms of its actual linear diameter in order to look as large as it does on the sky, in terms of its angular diameter.

Hubble had actually measured distances, of course, using the Cepheids and other techniques, in order to calibrate the redshift–distance relation. But the whole point of this exercise was to be able to use this relationship, once it had been calibrated, as a means of determining distances to other galaxies by measuring their redshifts. Redshift and distance are inextricably linked in the investigation of the distance scale of the Universe, and determinations of the true linear sizes of other galaxies. Hubble's high value of the constant that now bears his name goes hand in hand with a so-called short distance scale, and with the idea that other galaxies are so much smaller than the Milky Way that our own Galaxy has to be regarded as like a large continent, while the other spirals are no more than offshore islands.

This was an entirely plausible idea at the end of the 1920s, and most astronomers seem to have accepted it without a qualm. But, as I have mentioned, there was one notable exception — the pioneering astrophysicist and relativist Arthur Eddington.

Eddington always held the view that there is nothing special about the place that we human beings occupy in the Universe (much later, in the 1990s, this became known as the "principle of terrestrial mediocrity" — that we live in an ordinary part of the Universe). If there were anything in this idea, Eddington argued, then our home Galaxy could not be particularly remarkable. In his book *The Expanding Universe*, published as early as 1933, he wrote:

> The lesson of humility has so often been brought home to us in astronomy that we [he meant "I"] almost automatically adopt the view that our own galaxy is not specially distinguished — not more important in the scheme of nature than the millions of other island galaxies. But astronomical observation scarcely seems to bear this out. According to the present measurements the spiral nebulae, though bearing a general resemblance to our Milky Way system, are distinctly smaller. It has been said that if the spiral nebulae are islands, our own galaxy is a continent. I suppose that my humility has become a middle-class pride, for I rather dislike the imputation that we belong to the aristocracy of the universe. The earth is a middle-class planet,

not a giant like Jupiter, nor yet one of the smaller vermin like the minor planets. The sun is a middling sort of star, not a giant like Capella but well above the lowest classes. So it seems wrong that we should happen to belong to an altogether exceptional galaxy. Frankly I do not believe it; it would be too much of a coincidence. I think that this relation of the Milky Way to the other galaxies is a subject on which more light will be thrown by further observational research, and that ultimately we shall find that there are many galaxies of a size equal to and surpassing our own.

These are astonishingly prescient comments to have been made less than ten years after Hubble established once and for all that the spiral nebulae are indeed external galaxies, and within two years of the classic paper by Hubble and Humason on the redshift–distance relationship. But Eddington did not go so far as to suggest that the Hubble Constant should be reduced by a factor of ten (which is what would be required to make the distances to the galaxies studied by Hubble so great that the galaxies themselves would be about as large as the Milky Way), and one way in which the puzzle as he saw it could have been resolved would have been if bigger and better telescopes had found other spirals as big as the Milky Way beyond the swarm of lesser islands in our immediate vicinity.

As Eddington said, the key to a better understanding of the Universe at large was better observations. That meant new telescopes, bigger than the 100-inch Hooker Telescope that Hubble had used for his breakthrough work. And by 1933 the telescope that would take astronomers on their next step out into the Universe was already being planned for a new observatory being constructed on Mount

Palomar, a little to the south of Mount Wilson. It was the last of the great telescopes inspired by the work (and fund-raising skills) of George Ellery Hale, and although he did not live to see it become operational in 1948 (delayed, among other things, by World War Two), it bears his name. The Hale Telescope has a 200-inch diameter mirror, designed to be able to see galaxies a billion light-years away, increasing the volume of space visible to astronomers by a factor of eight, compared with the Hooker Telescope. Other things being equal, that would give the cosmologists eight times more galaxies to study (things didn't stay equal; since the 1940s, the range and sensitivity of the 200-inch have been greatly increased by the use of electronic detectors which monitor the light gathered by the mirror). Hubble was involved in the completion of the great telescope, and used it during the last few years of his life. But he was already in his fifties when the 200-inch was completed, and it would be his successor, Allan Sandage, who used the new instrument to make a revision to the cosmic distance scale that would surely have been to the satisfaction of Arthur Eddington.

In fact, the first step towards a revision of Hubble's distance scale had already been made by the time the Hale Telescope became operational, using the old Hooker Telescope. And it had been made, in no small measure, because of the way World War Two affected the astronomers on Mount Wilson.

The key player in this phase of the story was Walter Baade, a German-born astronomer who was just four years younger than Hubble, having been born in 1893. His major contribution to our understanding of galaxies and the distance scale of the Universe was actually made in two stages, over a period of almost ten years, using two different telescopes on two different mountains some 80 kilometres apart.

Unfortunately, the relationship between the two pieces of work has often been confused in previous accounts (including, I am ashamed to say, my own), but I will try to get it right here.

After working at the Bergedorf Observatory of Hamburg University, in 1931 Baade emigrated to the United States, where he joined the staff of the Mount Wilson Observatory just after Humason and Hubble had completed the first stage of their groundbreaking work on the redshift–distance relation. Baade was a fine astronomer, but had a sometimes disorganised private life, and didn't get around to obtaining the preliminary papers needed to become an American citizen until 1939. He soon lost these during a house move, and hadn't got around to renewing his application when the Japanese attacked Pearl Harbor in December 1941, and the United States declared war on Germany, as well as Japan.

There was an understandable (if somewhat paranoid) fear at the time that Los Angeles might be attacked from the sea, and Baade's status as an "enemy alien" in a militarily sensitive zone led to him being subject to a curfew, required to stay in his home each night from 8 P.M. to 6 A.M. — so he could do no observing. At the same time, other astronomers, including Hubble, were being recruited to help with the war effort, and there were few people around to use the telescopes on Mount Wilson. It took several months to persuade the authorities that Baade did not really pose a threat to the United States, and to accept that his original application for citizenship, made before the attack on Pearl Harbor, showed good faith. But eventually he got back to the mountain, where he had about as much time as he wanted on the 100-inch telescope, and also benefited not only from the wartime blackout of the city below, but from a new kind of more sensitive photographic plate that had just become available.

With these advantages, Baade was able to use the telescope to photograph objects even fainter than anything Hubble had observed. He set about photographing the Andromeda Galaxy, M31, in unprecedented detail, picking out individual stars where his predecessors had only been able to detect a fuzzy blur. And what he found was a spectacular tribute to his skill as an observer. Baade discovered that the Andromeda Galaxy (and, by implication, all spirals) is made up of two different kinds of star. The first kind, which he called Population I, occurs in the spiral arms of such a galaxy, and comprises mainly hot, young, blue stars, rich in heavy elements. The other kind, which he called Population II, occurs in the central part of a spiral galaxy (its nucleus) and in globular clusters. It consists of old, cool, red stars which contain very little material other than hydrogen and helium.

These differences are now explained in terms of the way galaxies and stars form and evolve. Population II stars actually came first, and are made of primordial material left over from the Big Bang; Population I stars are relatively young, and have been made out of material previously processed in at least one generation of earlier stars. Baade also discovered that there are also two types of variable star with Cepheid characteristics, one associated with each kind of stellar population. The Cepheid period–luminosity relation had been calibrated using Cepheids in our own neighbourhood, in a spiral arm of the Milky Way. They were Population I stars, and these Cepheids are now known as "Classical" Cepheids. The equivalent variables in Population II are now known as W Virginis stars. There is indeed an equivalent period–luminosity relation for these stars, but overall the W Virginis stars are fainter than the Classical Cepheids. Contrary to what you may read elsewhere, this told astronomers at the time (1944) nothing about the distance scale, because Hubble had indeed used the brighter Classical Cepheids in his work on the distance to the Andromeda Galaxy. But within a few years, Baade was able to transfer all the skills he had honed on the 100-inch on Mount Wilson, plus the new photographic plates, to the 200-inch on Mount Palomar, where he could look into the Andromeda Galaxy in even more detail, trying to photograph even fainter stars.

The stars that Baade was particularly eager to identify in the Andromeda Galaxy were the RR Lyrae variables, which we have already met. These stars are much fainter than Cepheids, but are very reliable distance indicators. At the distance Hubble had calculated for the Andromeda Galaxy, they ought to have been clearly detectable with the 200-inch using the techniques that Baade transferred from the 100-inch. But, to his initial consternation, they were not. RR Lyrae stars are what are now known as Population II objects, and are often found in globular clusters. The globular clusters could be seen in the Andromeda Galaxy, but not the individual RR Lyrae stars. Only the very brightest Population II stars could be resolved in the globular clusters in the Andromeda Galaxy, even using the 200-inch and all the tricks Baade had learned back at Mount Wilson. From studies of the globular clusters in our own Galaxy, astronomers already knew just how much brighter these brightest Population II red giant stars are than the RR Lyrae stars — in the units used by astronomers, the stars Baade could just resolve were 1.5 magnitudes brighter than the RR Lyrae stars he had expected to be just able to see. The only explanation was that the whole calibration of the distance scale was wrong by the same amount, and that the Classical Cepheids themselves were 1.5 magnitudes brighter than had been estimated, placing the Andromeda Galaxy much further away than had been thought.

The error in the calibration went right back to Shapley's work on the Cepheid distance scale, published back in 1919, thirty years before Baade's work with the 200-inch. It turned out that a combination of dust in the Milky Way and the fact that there are two kinds of Cepheids had conspired to fool Shapley, and everyone else at the time. Shapley had naturally used just about every Cepheid he had any information on in working out his distances and brightnesses to determine a period–luminosity relation, and some of these were (we now know) Population I Cepheids in the disk of the Milky Way, while others were Population II Cepheids in globular clusters. The Population I ("Classical") Cepheids are brighter, as Baade discovered in 1944. But there is more dust in the plane of the Milky Way, which makes them look fainter than they really are, while the Population II Cepheids (the W Virginis stars) are seen above and below the plane of the Milky Way, so they are less dimmed by dust. By an unfortunate coincidence, the dimming of the Classical Cepheids in Shapley's sample was just enough to make them match up in apparent brightness with the fainter W Virginis stars. Hubble had indeed been looking at Classical Cepheids in the Andromeda Galaxy — but the calibration he had been using for their brightnesses (the period–luminosity relation) was actually more or less the right calibration for the fainter W Virginis stars.

So the stars Hubble had used to estimate the distance to the Andromeda Galaxy were actually about twice as bright as he had thought. Since this measurement was the key to the way Hubble had worked his way out into the Universe, calibrating the brightnesses of other objects (such as exploding stars, or whole globular clusters) from what he thought was the distance to the Andromeda Galaxy, this meant that all the distances worked out by Hubble had (eventually) to be doubled — and that meant that the galaxies were all bigger than Hubble had thought, while the Universe was twice as old as Hubble's first estimate of the redshift–distance relation had implied.

Baade's discovery that the distance scale actually had to be doubled, placing the Andromeda Galaxy at a distance of about 600,000 parsecs (about 2 million light-years) instead of the 250,000 parsecs (800,000 light-years) determined by Hubble, was formally announced in 1952, at a meeting in Rome. It reduced the accepted value of the Hubble Constant from over 500 km/sec per Megaparsec to just 250 km/sec per Megaparsec (from now on, I will just quote the number for the constant, and leave the units as understood), pushing the established age of the Universe back from about 1.8 billion years to about 3.6 billion years, and made the Universe about as old as the age then known for the Earth. This revision to the distance scale of the Universe (and its age) was soon confirmed by observations (at the limit of the then-available telescope technology) of RR Lyrae variables in the Small Magellanic Cloud, which also turned out to be 1.5 magnitudes fainter (in apparent magnitude, that is) than anticipated, placing them correspondingly further away than the distance determined for the Cloud by the Shapley Cepheid calibration. Newspaper headlines made much of the fact that the size of the Universe had "doubled," since the recalibration of the Hubble Constant meant that all distances inferred from redshifts were twice as great as had been thought.

Since we now estimate that the oldest stars are well over ten billion years old, though, you might wonder why cosmologists were pleased, in 1952, to have pushed their estimate for the age out to only about four billion years (actually still slightly less than the estimated age of the Sun at that time). The new age for the Universe was more reassuring to astronomers in the early 1950s than you might think from hindsight, however, because at that time they did not know just how old the stars were. And as stellar astrophysicists began to come up with numbers higher than five billion years for the ages of stars, in the 1950s and into the 1960s, Hubble's successor was using the 200-inch to push back the known age of the Universe by a roughly comparable amount, with almost every improvement in technology and observing power leading to a further downward revision of the Hubble Constant.

Allan Sandage was born in 1926, after Hubble had shown that the spiral nebulae are external galaxies, but before Hubble and Humason had discovered the redshift–distance relation. He became interested in astronomy when he was nine, after looking through a friend's telescope, and a few years later he read Hubble's classic book *The Realm of the Nebulae,* in which the great new cosmological discoveries were described in language accessible even to a teenager. His education was interrupted by being drafted into the U.S. Navy for eighteen months at the end of World War Two, but he graduated from the University of Illinois in 1948. At Illinois, Sandage's interest in astronomy led him to join a small team of dedicated amateurs carrying out a sky survey organised by Bart Bok, of Harvard University. The survey involved photographing specific areas of the sky, developing the plates, and working out the magnitudes of all the stars captured in the photographs. It was tedious, painstaking work, but it was to open the way for Sandage to become Hubble's heir.

In 1948, Sandage applied to do research in physics at Caltech, as close as possible to the famous 100-inch telescope on Mount Wilson, and the new 200-inch, in the hope that he might later be able to join the astronomers. To his delight, however, that year Caltech started its first Ph.D. programme in astronomy, specifically to meet the expected demand for people to work with both the big telescopes, and Sandage was selected as one of the first five students enrolled for the programme. After a year learning the basics of astrophysics (mainly taught by Jesse Greenstein), in the summer of 1949 Sandage was chosen to assist Hubble in his latest project, an attempt to determine the ultimate fate of the Universe, from the curvature of space.

This project did not immediately involve Sandage with the great

200-inch Hale telescope, which like all the big telescopes has a very narrow field of view. It is fine for focusing in on faint objects in a small patch of the sky, but no use for photographing a broad region of the heavens on one plate. That kind of survey work is the prerogative of another kind of telescope, the Schmidt camera, invented by the Estonian Bernhard Schmidt in the 1930s. Palomar had a Schmidt camera with an aperture of 48 inches (1.2 m), which could photograph a region covering 40 square degrees on a single plate; for comparison, the field of view of a conventional telescope is typically just half a degree (30 minutes of arc) across. This Schmidt could photograph thousands and thousands of galaxies, and many plates from its surveying work were already available. Even without measuring the redshifts of every galaxy he could photograph with the 200-inch, Hubble reasoned that fainter galaxies ought to be further away from us, and this, together with the survey of nearby regions of the Universe made by the Palomar Schmidt, was his hoped-for key to unlocking the fate of the Universe.

If space were flat, in the way I have already described, then equal volumes of space centred on the observer ought to contain equal numbers of galaxies, on average. It is as if you stood in a forest on flat ground, and counted how many trees there were in successively larger circles of ground around yourself—you would expect to find twice as many trees in a circle with an area of 200 square metres as in a circle with an area of 100 square metres. But if space is curved (the equivalent to your forest either being on a hilltop or at the bottom of a valley), then you would find either more or less trees than expected as you looked further and further away.

Ultimately, Hubble's plan was to probe the curvature of the far reaches of the Universe in this way, using the 200-inch, and counting the number of galaxies in a particular volume of the Universe, rather than the number of trees in a particular area of forest. But before he could make the necessary comparisons, he needed an accurate count of the number of galaxies in our part of the Universe, as a benchmark against which his long-range counts could be compared. That was where Sandage, and the Schmidt plates, came in. With his experience of counting stars of different magnitudes for Bart Bok's survey, he was the obvious choice to be given the task of counting galaxies of different magnitudes on the Schmidt plates being obtained at Mount Palomar, to provide the key measurement of the number of galaxies in the local volume of the Universe. But just as Sandage was getting into his stride on the project, in July 1949 Hubble suffered a heart attack, and was banned from the mountain by his doctor while recuperating. The project got pushed to one side, and Sandage (with his fellow student Halton Arp) was sent to Mount Wilson to learn observing from Walter Baade, then the acknowledged expert. (Incidentally, the Universe turns out to be so nearly flat that in spite of heroic efforts by observers over the past half-century, including Sandage, it is still impossible to tell, from these number counts of galaxies alone, whether it is just open or just closed, or precisely flat.)

The work Sandage and Arp now became involved in was photographing globular clusters, and analysing the light from their individual stars. The two students started out using the 60-inch telescope, which only a little over thirty years before had been the best telescope in the world, but was now relegated to the status of a suitable instrument for beginners. Even so, it was still carrying out important research (as, indeed, it still is, in a modest way, today). Then, they progressed to the 100-inch. At each stage, Sandage proved to be a superb observer — and it was this work on globular clusters and stellar astrophysics that formed the basis of his doctoral thesis. It was Sandage who, in 1952, found, from his study of the globular cluster M3, the turn-off in the main sequence which soon became a key way to measure the ages of such clusters (see Chapter 2).

In 1950, alongside his doctoral work, Sandage became Hubble's assistant, carrying out the observing programme on the 200-inch that Hubble had devised, but which the great man could no longer complete on his own (although he was allowed to return to the mountain and do some observing from October 1950 onwards). These were exciting times on the mountain. Baade was carrying out the work that would lead to his dramatic revision of the distance scale; Humason, who had introduced Sandage to the 200-inch, but would himself be sixty in 1951, was breaking records with his redshift work on the telescope, recording redshifts corresponding to velocities greater than a fifth of the speed of light; and Sandage, a graduate student, was continuing Hubble's work, searching for variable stars in faint galaxies and trying to find new stepping-stones to give distances further out into the Universe than anyone had probed.

In 1952, when Sandage had almost completed his Ph.D. work (the degree was actually awarded in 1953), he was offered a post as assistant astronomer at the observatory. He accepted the position, but immediately took a year away from California to work with Martin Schwarzschild, in Princeton, on the implications for stellar evolution of the discovery of the main sequence turnoff in globular clusters. Using calculators that were little more than glorified mechanical adding machines, they were able to work out, for the first time, how a star like the Sun evolves during its lifetime on the main sequence, and to work out when a star with a particular mass starts to become a red giant. They found that the cutoff in M3 corresponded to an age of just over three billion years. But once a star leaves the main sequence, things get complicated — too complicated to be calculated using the "computers" available at the time. So that was as far as the work could go, and Sandage went back to California to take up his new post.

He had scarcely returned to California (expecting to continue his work on globular clusters and stellar evolution) when, in September 1953, Hubble suffered a stroke and died. His new programme of cosmological surveying had scarcely begun, and nobody had yet had time to assimilate Baade's revision of the distance scale. Sandage inherited the task of finishing the job Hubble had begun, even though he had expected to carry out research in stellar evolution. As he later told science historian Alan Lightman:

> I felt a tremendous responsibility to carry on with the distance–scale work. He had started that, and I was the observer and I knew every step of the process that he had laid out. It was clear that to exploit Walter Baade's discovery of the distance–scale error, it was going to take 15 or 20 years, and I knew at the time it was going to take that long. So, I said to myself, "This is what I have to do." If it wasn't me, it wasn't going to get done at that period of time. There was no other telescope; there were only 12 people using it, and none of them had been involved with this project. So I had to do it as a matter of responsibility.

During the 1950s, Sandage set about recalibrating every step in the chain of distances used by Hubble. Remember that these steps depend on things like determining the average brightness of the brightest stars, or brightest globular clusters, found in particular types of galaxy, the average brightness of the brightest galaxies found in clusters of galaxies, and so on. And all these observations are plagued by the need to correct for extinction of light by dust in space (not just dust in our Galaxy, but dust in the distant galaxies being investigated).

Every correction that Sandage applied reduced the value of the Hubble Constant. In 1958, Sandage discovered the biggest single error in the chain of arguments that had been used by Hubble (not Hubble's fault — he did the best he could with the 100-inch telescope, and Sandage was working with the 200-inch). A key step in all this work is determining the distance to a large cluster of galaxies in the direction of (but far beyond) the constellation Virgo, and therefore known as the Virgo Cluster. Galaxies such as the Andromeda Galaxy and the Magellanic Clouds are (together with the Milky Way) part of a small system of galaxies known as the Local Group, and although measuring distances to these objects is a crucial step in calibrating the brightnesses of things like Cepheid variables and supernovae, it doesn't tell us anything about the redshift–distance relation, because the members of the Local Group are held together as a unit by gravity (so, as we have seen, the Andromeda Galaxy is actually moving *towards* us, and its light shows a blueshift, not a redshift); it is redshifts relative to the Local Group as a whole that tell us something about the nature of the Universe at large.

The Virgo Cluster contains at least 2,500 identified galaxies, two-thirds of them spirals. It is both far enough away and rich enough in its variety of components for many of the more long-range distance estimators (such as the brightnesses of individual galaxies) to be calibrated by studying the cluster. But it turned out that some of the objects that Hubble had identified as individual bright stars in galaxies in the Virgo Cluster were actually large clouds of hot gas (known as HII regions), with several bright stars embedded in them. The HII regions are far brighter than individual stars, and in order to look as faint as they do to telescopes on Earth they have to be much further away than Hubble had thought. So the Virgo Cluster itself, the first step out into the Universe beyond the Local Group, is much further away than he had thought, and all distances measured relative to the Virgo Cluster have to be revised upwards accordingly — over and above Baade's revision of the distance scale, which had increased the distance to the Andromeda Galaxy, and thereby every distance determined relative to that galaxy, including the distance to the Virgo Cluster.

With all of the revisions Sandage made to the distance scale in the 1950s, but especially this discovery that HII regions had been misidentified as individual stars, by the end of that decade he had reduced his best estimate of the Hubble Constant to just 75, in the usual units. Being a scrupulously honest observer, Sandage also made a serious attempt to make allowance for the remaining uncertainties in the various techniques he had used to arrive at this number, and concluded that he could be wrong by as much as a factor of two. In other words, although his "best estimate" was 75, the "error bars" were so large that the number could be as small as 38 or as large as 150. The astronomer Virginia Trimble, who has made a special study of the history of the investigation of the Hubble Constant, said at the end of 1996 that this was "the last realistic set of error bars published for a very long time," and "the last completely non-controversial" value published until the present day.

The seeds of what was to become a long-running controversy about the size of the Hubble Constant and the age of the Universe were sown almost as soon as Sandage had published his honest assessment. Other astronomers, without access to the 200-inch but with the benefit of other technological improvements in the 1950s, were also able to make partial corrections to the distance scale determined by Hubble, but without including all of the factors that went into

Sandage's number. Everybody was convinced that their own number was the best, and different people had different ideas about how much to allow for effects like extinction. So, by the beginning of the 1960s, as well as Sandage's value of the Hubble Constant there was an estimate that the number lay in the range 143 to 227, another estimate of 125 ± 5, and one of 134 ± 6. With hindsight the error bars on these last two estimates look ludicrously small (and nobody except the authors of those two papers thought much of them at the time), but because Sandage had been so honest about his own error bars, it looked as if all these estimates could be made to overlap.

The weight of Sandage's authority and the power of the 200-inch did have some influence on the way opinion developed, though, and in the early 1960s most cosmologists began to use as a rule of thumb a value of H of 100.* This owed something to a natural tendency to average out the various estimates (natural, but not good science unless you have sound reasons for thinking that all the estimates are equally reliable), and something to the appeal of a "round number." It has always seemed to me, though, that there was a psychological barrier at 100, rather like the way products are priced at $99.95 instead of at $100, but in reverse. Hubble himself had come up with a three-figure number for H, 525. Reducing this number as low as 100 was worrying enough to cosmologists brought up thinking of Hubble as the Oracle; reducing it to two figures (even to 95) was, somehow, an even bigger psychological step — except for Hubble's heir.

That was still more or less the situation when I started studying astronomy in the mid-1960s. At that time, most cosmologists used

---

* Strictly speaking, in cosmology the symbol H denotes the Hubble Parameter, while the symbol H_0 is used to refer to the present-day value of the parameter, the Hubble Constant. I shall just use H for the Hubble Constant, since there is no possibility of confusion arising in this book.

H = 100 as a rule of thumb, partly because it is a nice round number to work with, but they no longer regarded this as a number set in stone, and would have been not too unhappy with a value of H = 50 (in effect, Sandage's estimate of H = 75 still stood, but with the error bars reduced to a range of ±25). Quite apart from the relevance of all this to the age of the Universe problem, this range of numbers is particularly interesting in the light of Eddington's prescient comments about the size of our own Galaxy. A number at the bottom end of this range would place all the other spiral galaxies at just the right distances for the Milky Way itself to be an average-sized spiral. A number near the top end of the range, however, would bring all the other galaxies that much closer to us, meaning that they must be systematically smaller than the Milky Way — and that the Milky Way is about twice as big as an average spiral.

From one perspective, this is much more worrying than the idea which troubled Eddington, that the Milky Way is a continent among islands. If the Milky Way really were the only huge galaxy in the Universe, it would be quite likely that we would be living in it, just as in Scotland any individual person chosen at random is more likely to be found living on the mainland than on one of the Scottish isles. If, however, the Milky Way were the largest galaxy around but only by a small amount, it seems rather odd that it should just happen to be our home. This is the line of reasoning known as the principle of terrestrial mediocrity, which says that there is nothing special about our place in the Universe. If there is anything in this argument, we ought to live in a more or less ordinary-sized galaxy — perhaps a bit larger or a bit smaller than average, but not the biggest or the smallest. This kind of reasoning, with cosmic inferences being drawn from simple logic, appealed to me as a naive student, and with Eddington to back me up I used to argue, whenever the question came up in discussions, that "of course" H must be closer to 50 than to 100. Nobody took much notice (in the small circle of astronomers I had contact with at the time), not least because there was no real debate about which end of the accepted range of values H might lie. It was just generally accepted (around 1966 or 1967) that observations would get better, the number would be pinned down more accurately, and that would be very nice, but nothing to get worked up about. But things didn't work out that way.

Although I drifted away from astronomical research and into scientific journalism at the end of the 1960s, I kept tabs on cosmological developments, and was pleased to see Sandage and his colleagues modestly reducing their own estimate of H, and considerably reducing their error bars — just the kind of quiet progress that had been anticipated. But I was totally baffled when, at the same time, a second group of astronomers began to favour the higher value of H (about 100), while also narrowing their own claimed error bars. By the late 1970s, there were two clearly opposed schools of thought, one arguing for a value of H close to 50, the other arguing for a value of H close to 100, and each claiming that their own error bars completely ruled out the other possibility. It was against this background that, over a period of more than twenty years, astronomers developed new techniques for measuring the Hubble Constant — some of which operated entirely independently of the traditional distance ladder based on Cepheids, others which still used Cepheids for measuring distances to the nearest galaxies, but then leaped out into the Universe at large in a single bound. And, along the way, as the dust cleared and honest error bars became fashionable once again, it became clearer than ever before that whatever the exact value of H, the general agreement of all these techniques was telling us something truly significant about the nature of the Universe we live in.

7

New Rulers

From Controversy to Consensus

The trouble with working out just how fast the Universe really is expanding — working out the value of the Hubble Constant — is that until you get a long way out into the Universe at large, local effects swamp the quantity you are trying to measure. It is no use trying to measure H from, for example, measuring the distance to the Andromeda Galaxy and comparing this with the redshift of the Andromeda Galaxy. In this case, the local movement of the Andromeda Galaxy through space under the gravitational influence of its neighbours in the Local Group of galaxies, combined with our own motion in orbit around the centre of the Milky Way, means that we actually see a blueshift in the light from that galaxy, not a redshift. The only value of the Andromeda Galaxy in determining H, as we have seen, is that because we know its distance very accurately (from Cepheid measurements), we can calibrate the brightnesses of things like HII regions and globular clusters from studying that galaxy.

Things are almost as bad for the Virgo Cluster of galaxies, the next key stepping-stone out into the Universe at large. The first problem is the size of the cluster, compared to its distance from us — indeed, it is so big that it is hard to say what its distance from us is. The Virgo Cluster is like a huge swarm of bees, all moving around relative to each other (relative to the centre of the swarm), but with the swarm as a whole moving through the air, just as the cluster as a whole is being carried along by the expansion of space. When we try to measure the distance to the Virgo Cluster, what we do is like measuring the distance to some of the individual bees (galaxies). Unfortunately, the swarm is not the only thing we have to worry about in the Universe — there are stray "bees" between us and the swarm, and other strays beyond the swarm, along the line of sight. So even if you have measured the distance to an individual galaxy accurately, it is hard to be sure if that particular bee really is a member of the swarm.

There is another complication, because it is easier to measure distances to galaxies that are closer to us. So it is easier to measure distances to galaxies on our side of the Virgo Cluster, and unless you take great care about how you interpret these measurements and take averages it will look as if the cluster is a lot closer to us than it really is. In fact, the shortest measured distances to galaxies identified as members of the Virgo Cluster are all about 17 Megaparsecs. The greatest distance measured to any galaxy firmly identified as belonging to the cluster is around 25 Megaparsecs. From this, and other evidence, the centre of the cluster probably lies about 21 Mpc away from us (but don't take that number as gospel, even today; people still argue about it).

If we accept 21 Mpc as "the distance" to the Virgo Cluster, and guess that that is the distance to any member of the cluster whose distance we cannot measure directly, then we are immediately introducing a possible error of ± 4 Mpc into our calculations. At a distance of 21 Mpc, that is an error of just under 20 percent. So when we try to calibrate the distance indicators (such as the average brightnesses of entire galaxies) by using statistics from the Virgo Cluster, we already have an error that big lurking in our equations.

The problem is not that the Virgo Cluster is so big, but that it is so close. Of course, it *has* to be close, in order for us to measure the distances to any of its galaxies using the relatively short-range techniques calibrated by studying the Andromeda Galaxy (the same sort of approach has been used with other clusters of galaxies, but is inevitably plagued with the same problems). If a cluster the same size as the Virgo Cluster (8 Mpc across) were at a distance of 100 Mpc from us, then the uncertainty in our distances to the individual galaxies would still be ±4 Mpc, but that would now only represent an error of 4 percent, not 20 percent. Except that we wouldn't have any distance measurements, because a cluster at 100 Mpc from us would be too far away for any of the distance indicators calibrated in the Andromeda Galaxy to be visible!

There's another problem with using the Virgo Cluster to determine H directly. The redshifts of the individual galaxies in the cluster are an unreliable indication of the cosmological redshift. First, there is the difficulty that the individual galaxies are moving around, relative to one another and the centre of the cluster, just as in the Local Group the Andromeda Galaxy is moving fairly rapidly relative to us. Indeed, one of the few Virgo galaxies that has a very good distance measurement (from supernova observations) has a distance of 25 Mpc (on the far side of the cluster) but a very low redshift, presumably because it is falling in towards the centre of the cluster, and therefore moving towards us, which cancels out part of its cosmological redshift. So to get a meaningful redshift for the whole cluster you would have to measure very many redshifts for individual galaxies (not too difficult) and take an average — but without knowing the distances to each individual galaxy, we can't be sure that we aren't measuring the redshifts of all the galaxies on our side of the cluster, and getting a completely biased result.

There's more. Because we are so close to the Virgo Cluster, the motion of the Milky Way (and the whole Local Group) is influenced by its gravitational pull. We are indeed being carried away from the Virgo Cluster by the expansion of space, so that the light from all the galaxies in the cluster does show a redshift. But at the same time we are falling towards the Virgo Cluster through space, because of its gravitational pull. The best analogy is with someone trying to walk down a very long escalator that is moving upwards. The Local Group is the person walking down the escalator, towards the Virgo Cluster, which is at the bottom of the escalator. We really are moving downwards relative to the steps on the escalator, as we walk down them. But the escalator is carrying us backwards faster than we are walking, so the distance between us and the bottom of the escalator is increasing, although not as fast as it would be if we were not walking downwards. Astronomers still argue about just how big this "Virgocentric infall" is, but the argument covers the range from about 200 kilometres per second to 300 km/sec. The best estimate of the average redshift for the Virgo Cluster (with all the caveats I have already mentioned) corresponds to a velocity of about 1,000 km/sec relative to the Local Group of galaxies, so the inference is that the cosmological redshift (which is not a Doppler effect, remember, but is caused by the stretching of space) of the cluster is 1,200–1,300 km/sec, with about 25 percent of this being cancelled out by the true Doppler effect of our infall motion — we really are falling towards the Virgo Cluster even though we are actually getting further away from the cluster as space expands.

With all of this uncertainty, what use *is* the Virgo Cluster in determining H? Like the Andromeda Galaxy, its value lies in providing a way to calibrate the brightnesses of things that can be used as stepping-stones to take us further out into the Universe. With so many galaxies in the Virgo Cluster, for example, it is even possible to get some idea of the average brightness of particular kinds of galaxy. Then, you can look for the same kinds of galaxy in much more distant clusters, and compare their brightnesses with the brightnesses of the same sort of galaxies in the Virgo Cluster. And *then,* you can get further out into the Universe still by using the brightnesses of whole clusters of galaxies as (admittedly rough-and-ready) distance indicators.

The technique gives you the distances to the more distant clusters in terms of the distance to the Virgo Cluster — we can infer that a more distant cluster is five times further away, or ten times further away, or whatever, than the Virgo Clusters. We are still left with the uncertainty introduced into our calculations by not knowing the exact distance to the centre of the Virgo Cluster, but now we can reasonably hope that the more distant cluster will be so far away that its cosmological redshift will completely overwhelm any small corrections needed to allow for the motions of its individual member galaxies, and ourselves, through space. A correction (or potential error) of a couple of hundred kilometres per second is something to worry about when we are dealing with a cosmological redshift of just over 1,000 kilometres per second — it represents an error of 20 percent. But for a cluster ten times further away than the Virgo Cluster, with a cosmological redshift of some 10,000 km/sec, the same sort of random velocities would introduce an error of only 2 percent, not 20 percent, in the final determination of H.

I don't intend to take you through all of the steps in the argument used to determine the Hubble Constant in practice, but I hope I have given you a flavour of the difficulties which plagued the measurements until very recently. It was these difficulties that led to the emergence of two camps in the 1970s, one arguing for a value of H around 100, the other arguing for a value close to 50. Since it is now clear which school of thought was correct, I won't go into all the details of an argument which seemed important to its protagonists, but is unlikely to be accorded much space in the astronomical history books. Instead, I want to tell you a little bit about *why* the argument arose, because it highlights the difficulties facing cosmologists when they try to make any accurate measurements of the properties of the Universe.

By the end of the 1950s, Humason and Baade had both retired (leaving a legacy of hundreds of photographic plates of distant galaxies) and Allan Sandage was the only astronomer left on Mount Palomar still pursuing the elusive Hubble Constant. The task of analysing all the plates, searching for Cepheid variables and so on, was too much for one person, and in any case Sandage had other interests (not least what proved to be the even more elusive search for a definitive measurement of the curvature of the Universe). In 1962, he recruited the Swiss astronomer Gustav Tammann, who was then thirty, but had not long completed his Ph.D. (because of a diversion into law when he was younger), to help. Tammann arrived at Caltech in February 1963, and set to work on the tedious task of comparing plates of the same galaxies taken at different times, searching for variable stars, particularly Cepheids and novae.

Using mounds of data, studying faint stars that were barely detectable in the photographic plates, using every calibrator they could lay their hands on, and making the best estimates they could for the effects of interstellar extinction, Sandage and Tammann edged their way out into the Universe. One of their key measurements was the

distance to the nearest really large spiral galaxy, a giant known as
M101 (it happens to lie roughly in the direction of, but far beyond,
the constellation known as the Plough). They suspected that all giant
spirals like M101 must be about the same brightness, and that if they
could measure the distance to M101 and thereby determine its true
brightness, they could use the apparent brightnesses of similar giant
spirals as a measure of their distances. They analysed everything they
could in M101 itself, and in smaller satellite galaxies in orbit around
the giant. From all the evidence, by the end of the 1960s they had a
distance of 7 Megaparsecs for M101, a key step in their determination
of a value close to 50 for the Hubble Constant itself.

The attack on this line of work, which went all the way back to
Hubble himself in the 1920s, came in the mid-1970s from Gérard de
Vaucouleurs, born in France in 1918, but by then working at the
University of Texas, Austin. In fact, de Vaucouleurs had been sniping
at Sandage's work for some time before they came into a full-blown
confrontation. His main interest was in the way galaxies are distributed across the Universe, and he was one of the first astronomers to
appreciate that they are distributed, not entirely uniformly, but in
superclusters and great sheets, wrapped around regions devoid of
galaxies, so that the overall appearance of the Universe is like a froth
of bubbles. One of the implications of this work is that there might be
large-scale effects of this distribution of galaxies on the way they
move, with the gravity of a large supercluster tugging on other galaxies and sending them streaming across the Universe. According to
this interpretation of the evidence, these streaming motions would
distort the appearance of the cosmological expansion, and lead to
incorrect estimates of the size of the Hubble Constant. In particular,
de Vaucouleurs suggested that the gravity of the Virgo Cluster was a

major influence which had fooled Sandage and Tammann into measuring a much lower value for H than applied in the Universe at large, beyond the region of influence of the Virgo Cluster.

But it wasn't just the streaming motions that worried de Vaucouleurs. As he got more involved in this aspect of cosmology, he convinced himself that just about every correction that Sandage and Tammann had made to the raw data in their assault on the Hubble Constant was wrong — that they hadn't allowed correctly not just for streaming motions but also for interstellar extinction, for example, or any of the other things that had to be allowed for. In September 1976, he said as much at a meeting held in Grenoble, in France. De Vaucouleurs told the meeting that the true value of the Hubble Constant was 100, and that, therefore, the Universe was only half as old (among other things) as Sandage and Tammann had estimated.

The argument raged, often bitterly, for twenty years. We now know that Sandage and Tammann were right all along, but that doesn't mean that the debate was not important. In particular, the fact that there was a debate encouraged the observers to come up with new techniques for measuring H, and it is in no small measure thanks to those new techniques (which might not have been developed so quickly without de Vaucouleurs' nagging) that we now know who was right. But *why* was de Vaucouleurs wrong? I asked Tammann that question in the spring of 1995, and he explained that it largely came down to something called Malmquist Bias (named after the Swedish astronomer Gunnar Malmquist, who drew attention to the problem in the context of stellar astronomy, in the 1920s), by which the average brightness of a group of distant objects (such as galaxies) seems greater the further away they are, because the faintest members of the group are too faint to be seen and taken into account. The same sort of argument, Tammann points out, also applies to the physical sizes of objects far away across the Universe — we can't see the smallest ones, so the average size we work out from the ones we can see is always too big.

In Tammann's words, the dichotomy in estimates of the Hubble Constant from the mid-1970s to the mid-1990s can be seen in terms of the different approaches of "optimists" and "pessimists." The optimists believe, he said, that their distance indicators are nearly perfect, with very little "scatter" (variation in properties such as brightness or size compared with the average). That means that they don't need to take account of the Malmquist effect, and "they blindly believe their distances which are derived from some relation which is locally calibrated. The price is that they have typically a [value of] H increasing with distance."

The pessimists, on the other hand, "hold that most distance indicators are quite lousy, i.e. they have important intrinsic scatter." This means that the further out into the Universe you look, the more your catalogues of distant objects will be biased in favour of larger and brighter objects, and the smaller and dimmer objects will be under-represented. "The obvious effect is that the mean luminosity of the catalogue objects increases with distance. An optimist, who denies the scatter, does not accept this increase and derives systematically smaller distances (and larger values of H)." In a final twist to his story, Tammann pointed out that several different distance indicators that have been used over the years all turn out to have about the same scatter, so they are all equally biased by the Malmquist effect, and give a consistently wrong answer — a reminder (like Shapley's original calibration of the Cepheid distance scale) that "consistent" doesn't necessarily mean "right." And in an aside dear to my heart, Tammann also pointed out (as if I hadn't known!) that "if H is bigger than 70, we have to accept that the diameter of our Galaxy and our neighbour

M31 [the Andromeda Galaxy] are larger than that of any spiral in the Virgo Cluster."

The other corollary of having the lower value of H is that there is no need to invoke large streaming velocities. This is not a problem, because there is a great deal of evidence (see *In Search of the Big Bang*) that although bright galaxies are distributed in a frothy way across the Universe, there is a great deal of unseen dark matter, even in the voids. So the distribution of mass in the Universe is more uniform than the distribution of bright galaxies would suggest, which means that the gravitational gradients are not so extreme, and the streaming motions are not so large as observations of bright galaxies alone might indicate. There is nothing inconsistent in adopting the pessimistic approach to the problem.

If you do that, it is possible to correct for Malmquist Bias using statistical techniques based on analysis of the measurements you can make of the distribution of brightnesses, or sizes, using the population of objects you are studying. Almost exactly twenty years after de Vaucouleurs set the hare running, Tammann summed up the pessimists' position at a meeting held in Baltimore, Maryland, in May 1996. Using the technique of measuring distances relative to the Virgo Cluster (to remove any influence from our infall towards the Virgo Cluster) he came up with a value for H of 54 ± 4 km/sec per Megaparsec. But even at the same meeting, other astronomers were still arguing the case for a value of H above 70, partly in the light of studies of Cepheids, made with the Hubble Space Telescope, in a few of the Virgo Cluster galaxies (to put the HST in perspective, it has a mirror about the same size as the one on the Hooker telescope, but the benefit of modern electronic detectors and the clear seeing possible from above the Earth's atmosphere).

As I shall explain in the next chapter, it was this early work on the

Virgo Cluster by the HST which triggered my own contribution to the debate. But hold on to that number, 54 ± 4, which represents the culmination of Edwin Hubble's own approach to the problem, building the cosmic distance ladder step by step outwards into the Universe from the Hyades Cluster in our own Milky Way Galaxy, to the globular clusters, the Magellanic Clouds, the Andromeda Galaxy, the Virgo Cluster, and beyond. Before I discuss in detail the work of the team I have been associated with, I want to tell you about some of the newer techniques which have brought home to cosmologists just how good the basic Einstein–de Sitter model of the Universe is.

One of these techniques still uses Cepheid variables, but gets out to the Universe at large with just one step, simplifying the whole business. Two other really good (that is, soundly based in well-understood physics) techniques each actually provide a way to determine H directly, without any need to calibrate distances using Cepheid variables. As yet, they are both uncertain techniques, in the sense that their estimates still have large error bars on them. But they definitely work. Then there is a fourth technique, which I shall mention briefly but which I have some doubts about, not least because it is not soundly based on thoroughly understood physics.

The first of these techniques is the one I have already mentioned in passing, using the brightnesses of supernova explosions in distant galaxies as an indicator of their distances. The key to this technique, of course, is being able to say with confidence that all the supernovae you are using have the same intrinsic brightness, and that you know what that brightness is. Then, you can measure distances as easily as you could measure the distance to a 100-watt bulb at the end of the street by measuring its apparent brightness. Until relatively recently, there was an element of wishful thinking and guesswork in this approach. But first astronomers identified a particular class of supernovae (known as

Type 1a) which really do all have the same maximum brightness, then they were able to pin down the distances to several galaxies in which these events had been seen from the Cepheid technique (in no small measure using the Hubble Space Telescope). Alongside all this, they developed a thorough physical understanding of what goes on in a Type 1a supernova (the different types are distinguished by how quickly they brighten and fade, their colours, the lines in their spectra, and so on).

All supernovae occur when a more or less ordinary star collapses to form a neutron star, a ball of material only about 10 kilometres across but containing slightly more matter than the Sun. This material is literally at the same density as an atomic nucleus, and a thimbleful of the stuff would (if it could be transported to Earth and magically prevented from expanding) weigh as much as all the human beings on the planet put together.

The energy released in a supernova explosion is essentially gravitational energy. If you start with a cloud of gas in space and let it shrink under its own weight, it gets hot inside as gravitational energy is released. That is how stars form in the first place. The cloud of gas gets so hot inside that nuclear reactions begin, generating heat which stabilises it and stops it from shrinking further. As we have seen, a star stays much the same size during its lifetime by burning nuclear fuel, converting hydrogen into helium and helium into heavier elements. But when it runs out of fuel, it has to collapse further. If it is heavy enough (at least several times as massive as the Sun at the end of its life) its core will collapse all the way to a neutron star (or possibly even a black hole), becoming a supernova which blasts its outer layers away into space, and briefly shines as brightly as a whole galaxy of main sequence stars. This is not, though, the kind of supernova we are interested in here.

If a star reaches the end of its life with about as much mass as our Sun, it will not be heavy enough to collapse all the way down to a neutron star. Instead, it will end its life as a white dwarf, a star with roughly the same mass as the Sun in a ball about as big as the Earth. On its own, such a star will simply sit quietly in that state forever, gradually cooling as it radiates away the last of its heat. But most stars are not on their own — they come in pairs, or more complicated systems. A white dwarf star in a binary system can gradually strip material away from its companion star, as a result of tidal forces (especially after the companion has left the main sequence and become a red giant). The stream of material from the companion star onto the white dwarf will slowly increase its mass, until at a crucial point something has to give.

The crucial point is known as the Chandrasekhar limit (or Chandrasekhar mass), and it is the maximum mass a white dwarf star can have without collapsing to become a neutron star. It is a very precise limit, 1.4 times the mass of our Sun, and what happens is very well understood in terms of the equations of quantum physics. That is what makes a Type 1a supernova — a white dwarf star with less mass than the Chandrasekhar limit gradually accumulating mass from a companion star until the critical point is reached and it collapses, releasing a blast of energy.

The beauty of it is that it doesn't matter what mass the white dwarf was to start with. At the point where it explodes, it has exactly the Chandrasekhar mass. So all Type 1a supernovae explode in the same way, and release the same amount of energy. They are all the same brightness. This is what I mean when I say that this distance indicator is based on sound physics.

Until the advent of the HST, though, even though the apparent brightnesses of supernovae in distant galaxies could be measured, there were very few direct measurements of the distances to any of the host galaxies, using the Cepheid technique (the HST was launched in 1990, but was no use for this kind of work until a fault in its optics was corrected in December 1993). And, whatever the theory said, the technique could only really be accepted as reliable if the distances to several such host galaxies could be measured, to prove that the Type 1a supernovae really did all have the same brightness at peak luminosity.

To give you some idea of how crucial every drop of information was (and to highlight the value of keeping good records), as recently as 1995 an important contribution to the search for the Hubble Constant was made by a researcher analysing a series of photographic plates taken exactly a hundred years earlier, in 1895, when Hubble himself was only six years old.

The plates showed the brightening and subsequent fading of a supernova, now identified (from the evidence on these plates) as a Type 1a supernova, in the galaxy NGC 5253. They became important in the mid-1990s because in 1994 Allan Sandage and his collaborators reported that they had identified Cepheids in NGC 5253 (using HST data), and obtained a distance to the galaxy. So Bradley Schaefer, of Yale University, went back to the original photographic plates (almost a hundred years old) and scanned them using modern technology to get an accurate measurement of the apparent brightness of the supernova, calibrating it against the brightness of the stars on the plates, which (unlike the supernova) can still be seen and analysed today. The best (most likely) value Schaefer found for the Hubble Constant from this single supernova was 51 ± 7, although because there were only a few plates showing the 1895 supernova, he did not have a complete record of its rise and fall, so there is not enough evidence to completely rule out slightly higher or lower values. Schaefer published his result exactly one hundred years after the supernova had reached maximum brightness.

One swallow doesn't make a summer, and nobody would trust a hundred-year-old photographic plate on its own to tell us the age of the Universe. But in the couple of years following Schaefer's analysis the supernova data began to build up, all pointing in the same direction. Part of the power of the technique is that these supernovae can be seen at distances not just of a few Megaparsecs, but well beyond 1 billion parsecs, so that once they are calibrated they provide a probe of the far reaches of the Universe, and open up, once again, the possibility of measuring the curvature of space (early results from this work suggest that the Universe is just open, but the jury is still out).

The supernova results that have come in since 1995 all point to much the same conclusion. In 1996, Sandage's team reported another analysis suggesting a value of H of 57 ± 4, and the same year David Branch and colleagues at the University of Oklahoma used HST data for the Cepheid distances to host galaxies for such Type 1a supernovae to calibrate the supernova distance scale and come up with a figure of 57 ± 5. Branch told me that "nothing over 70 is possible in the supernova world view," and said that "if anything I worry that a number around 60 may be too high, not too low." A review by Schaefer of ten Type 1a events, also published in 1996, concluded that $H = 55 \pm 3$.

But I was particularly interested in Branch's result at the time (and still am) because a year earlier he had also been involved in another attack on the problem, which got around the need to use Cepheids as distance indicators. The theoretical astrophysicists worked out an equation which related both the maximum brightness of a supernova and the time it takes to reach maximum brightness to the energy going into the explosion. As you might expect, a small explosion quickly reaches a feeble maximum, while a big explosion takes longer to reach a higher maximum. Applying this equation to the measured "rise times" for four Type 1a supernovae, the Oklahoma team came up with a value for H of 50, with rather large uncertainties ($+12$; -10), meaning that they couldn't rule out any value from 40 to 62 (the uncertainties come from the observations, not the theory; it is rare to catch the very beginning of a supernova on a photograph, so the measured rise times are not precise). But not only does this number agree with all the other supernova measurements (and the traditional measurements of Sandage and Tammann), it gets there without using Cepheids at all.

I've made the point, but for completeness I will just mention the latest results that were to hand in the spring of 1998, obtained by researchers at the Lyon Observatory. The importance of this study is that it used data from the Hipparcos satellite (more of this in the next chapter), which has obtained more precise geometrical parallax distances for stars in our own Galaxy, leading to a slightly more accurate calibration of the Cepheid distance scale. Using this improved distance scale to compute the distances to four galaxies in which Type 1a supernovae have been seen (including NGC 5253), the French team came up with a value for the Hubble Constant of 50 ± 3. And although the error bars on this number may be a little optimistic, you will see in Chapter 8 why I was keen to include this latest assessment of the supernova distance scale.

But before I get on to my own work, I want to pick up that point from the 1995 study by the Oklahoma team — that it is now possible to measure cosmological distances without using Cepheids at all. These techniques are in their infancy, which is one reason why some of them have large error bars. But at least those error bars all overlap with one another, and with the results obtained by what might now be called classical techniques. This is telling us something very profound about the Universe and our understanding of it.

One of the key predictions of the general theory of relativity is the way that light gets bent when it passes by a massive object. Isaac Newton's theory of gravity does also predict a light-bending effect, but not as big as the light-bending effect predicted by Einstein's theory. Indeed, it was measurements of the bending of light passing close by the Sun from distant stars, observed during an eclipse in 1919, that confirmed the accuracy of the general theory of relativity, and made Einstein famous. Einstein himself realised that under the right circumstances this light bending could act as a kind of gravitational lens, with a concentration of mass along the line of sight to a very remote object focusing light from that object and making it visible to telescopes on Earth.

Einstein wrote about gravitational lensing, as a theoretical prediction, in the 1930s. In the middle of the 1960s, astronomers realised that if they ever did find gravitational lensing at work in the Universe, it ought to be possible to use the images produced by the lensing effect to determine the Hubble Constant. But it was only in 1979 that the first clear-cut example of a double image on the sky produced by gravitational lensing was identified — and it took a further seventeen years of observation and theoretical work before the astronomers were able to use the data from this object, a quasar known as 0957 + 561, to get a reliable determination of H.

The procedure which is used to pluck a value for H out of the observations of the multiple images is very straightforward mathematically, but involves some algebraic and geometric details that I won't go into here. What matters is that for strong examples of gravitational lensing, the effect usually produces either two or four images of the distant object, because light has travelled from the object to our telescopes by two or four different routes around the intervening mass. Because the light that comes one way around has travelled a different distance from the light that comes the other way, it takes a different amount of time to reach us. The amount of time it takes depends on the geometry of the situation, the amount of mass in the lens (and the way it is distributed) and the actual distance to the distant object.

The light that forms one image takes a certain amount of time to reach us, and the light forming the other image takes a different amount of time to reach us. Each journey time depends on the distance *along each path,* which depends on the lens properties and on the redshift of the distant object (which can be measured) and Hubble's Constant (it also depends slightly on the curvature of the Universe, but only slightly). You also need to know the redshift of the object that is doing the lensing, which is usually a whole galaxy, visible between the images of the more distant object, but actually much closer to us along the line of sight. By comparing the brightnesses of the two images (which also depend on the lens properties), the redshifts and the time difference, it is possible to cancel everything else out of the equations, and be left with a value for H.

There are two problems with this. The essential first step in the whole process is to measure the time difference for light travelling along two different paths around a gravitational lens. In order to do this, the observer would like to see a sudden, clean-cut change in one of the images — a short-lived brightening or flare, perhaps, or a sudden dimming. Then, they start counting the days until the other image shows exactly the same pattern of flaring (or fading) activity. The difference is the key time delay needed in the calculation (typically, the time delays involved are a few tens or a few hundreds of days). Unfortunately, the best candidates for this kind of work are quasars, which are very energetic, very distant objects with high red-shifts, but which are each concentrated in a small volume, so they show up as starlike points of light on the sky (quasars are probably the very active central cores of young galaxies, powered by supermassive black holes).

The way you identify a gravitational lens at work is if you find two (or more) quasars which are very close together on the sky, and have exactly the same spectra and redshift, showing that they are really different images of the same object. But the unfortunate thing is that most quasars are fairly well-behaved beasts, not prone to sudden, dramatic changes in brightness. In the case of 0957+561, in spite of several false alarms it was not until 1995–96 that a sudden change in the brightness of one image was followed unambiguously by the same change in the other image 417 days later (more than a year later; the length of the time delay is another reason why these investigations take so long). The accepted value for the time delay in this particular system is now 417 ± 3 days.

This is where the theorists come in. It would be easy to turn this time difference into a value for H if we knew exactly how the mass in the gravitational lens is distributed (a whole cluster of galaxies is involved in the lens itself, although the gravity of one particular galaxy dominates the effect). If all the mass of the lens were concentrated in one lump, for example, we would get a certain value for H; but if the same amount of mass were distributed in two (or more) lumps, we would get a different value for H for each mass distribution. The theorists can calculate several different mass distributions which each match the optical observations of the lens system. At this point, though, the radio astronomers come into the story. They obtained radio maps of 0957+561 which showed that the quasar has a jet of material coming out of it, with five distinct "knots" in the jet. The same five knots show up in each of the two images, and by comparing the brightnesses of the individual knots in each image the theorists were able to work out a single best model of how the mass that is causing the lensing effect is distributed. Even then, there is still some uncertainty about the mass distribution, and this is the main cause of the remaining error bars on the number that emerges. Several different teams tackled the problem in 1997, coming up with slightly different numbers and slightly different estimates of the errors, but all in the same ballpark. One team found a value for H of 64 ± 15, one a value of 63 ± 12, and another 62 ± 7, each assuming a flat Universe (if the Universe is actually just open, these numbers should be reduced a little). The third value is probably genuinely more accurate, because it uses spectroscopic data from the Keck telescope on Hawaii, at the time the most powerful ground-based optical telescope in the world, with a mirror 10 metres (some 400 inches) across, although the very latest estimate claims to reduce the uncertainties to 59 ± 3.5.

As the first clear-cut result from the gravitational lensing technique, this is impressive, and in comfortable agreement (allowing for the error bars) with the other values of H we have discussed. But in order to be absolutely sure that the technique works, astronomers would like to have a minimum of four different gravitationally lensed systems, all giving the same answer. As yet, they have only two — $0957+561$, and a quadruple system known as $PG1115+080$ (the numbers in all these "names," by the way, correspond to the positions of the objects on the sky, cosmic latitude and longitude; $PG1115+080$ was the second gravitationally lensed system to be identified, in 1980).

In some ways, a quadruple system is better for the cosmologists. Instead of comparing brightnesses and time delays for just one pair of images, they can compare each of the images with the three others.

But the value of H that you get from quadruple systems is even more sensitive to the way the mass in the lens itself is distributed, so the modelling is even more important. At the end of 1996, astronomers found that a change in brightness of one of the components of this particular quadruple system was followed by a corresponding change in two of the other components about nine and a half days later, and in the fourth component twenty-four days later. The interpretation of these measurements depends crucially on the modelling, but when the result was announced it caused a brief flurry of mirth in some press reports, because the simplest model implied a value for H of 42 (with error bars of $\pm$ 6). The temptation to pick up on Douglas Adams' *Hitchhiker's Guide to the Galaxy* and report that "the answer to the Universe is 42" proved irresistible in some quarters, even though a closer reading of the research paper showed that a more realistic mass distribution for the lens suggested a value of 64 $\pm$ 9, while a value as high as 84 $\pm$ 12 could not be ruled out (the errors are bigger for the bigger values of H because in this case they are all the same percentage, about 14 percent).

The investigation was taken up by a team at the Harvard–Smithsonian Center for Astrophysics, in Cambridge, Massachusetts, who first suggested a value of the Hubble Constant of 60 $\pm$ 17 (based on improved modelling of the same quadruple lens system), and then, in the summer of 1997, refined their models still further to come up with a value of 51 $\pm$ 13, which still (in the spring of 1998) looks the best and most honest assessment of the situation regarding PG1115+080, although another slightly different model gives a value of 53 $\pm$ 9.

There are only two other systems in which even preliminary time delay measurements have been reported, and although about forty multiply imaged quasar systems are now known, most are too faint to make good candidates for this kind of investigation (too faint for redshifts to be measured in some of the components, maybe, or too faint for the flickering of the quasar images to be reliably recorded, or both). But it seems we may be on the edge of establishing the technique as a tool of comparable value to the traditional techniques and the supernova technique for measuring H. The next technique I want to tell you about is frustratingly difficult to work with at present, but offers the best hope of all of getting a truly cosmic perspective on how fast the Universe is expanding.

One of the most important of all cosmological observations, after the discovery that the Universe is expanding, was the discovery, in the 1960s, of the weak hiss of radio noise that fills the Universe and is now known as the cosmic microwave background radiation. I have explained the full significance of this discovery in my book *In Search of the Big Bang;* what matters here is that this radiation is interpreted as the leftover heat from the cosmic fireball in which the Universe was born, the Big Bang itself. As the Universe has expanded, this radiation has been redshifted and cooled until today it has a temperature only 2.7 degrees above the absolute zero of temperature, corresponding to *minus* 270.3 degrees on the familiar Celsius scale. The radiation comes from all directions on the sky, and was predicted by models of the Big Bang; indeed, its discovery was the clinching evidence that persuaded many astronomers and physicists that there really was a Big Bang. Even detecting this radio noise (equivalent to the radiation of a very, very cold microwave oven) is an impressive achievement, but radio astronomers are now able to detect tiny differences in the strength of the radiation (that is, tiny differences in its temperature) from different patches of sky, and use this to determine the value of the Hubble Constant, and thereby the age of the Universe.

The technique is based upon something called the Sunyaev–Zel'dovich effect, after the two astronomers, Rashid Sunyaev and Yakov Zel'dovich, who predicted it in the early 1970s, long before radio astronomy techniques were subtle enough to measure the effect. It is usually referred to in shorthand as the S–Z effect. What happens is that when the background radiation passes through a cluster of galaxies, the hot gas in the cluster (in between the galaxies) interacts with the photons that make up the background radiation and gives them a small boost in energy. This gas has a temperature of several hundred million degrees. The boost in energy it gives to the photons corresponds to shifting the photons to shorter wavelengths. Although this means that, overall, the radiation passing through the cluster has got a little bit hotter, it just happens that the photons given a boost in this way have been taken away from the region of the spectrum where the radio telescopes are sensitive, so as far as the radio observations are concerned they have been lost, and in that part of the spectrum the radiation look a tiny bit cooler, not hotter. The effect is very small indeed — about one part in ten thousand, or 0.01 percent, change in a temperature that is itself less than three degrees absolute — but it has now been measured for a few clusters. The cosmic microwave background radiation really is that tiny bit cooler, at radio wavelengths, in the direction of clusters of galaxies than it is from the rest of the sky.

This is an important discovery in itself, because it confirms, just in case anyone was still in doubt, that the background radiation really is just that — *background* radiation that comes from very far away across the Universe, beyond the clusters of galaxies. This is welcome additional confirmation that it originated in the Big Bang. But how can the S–Z effect tell you the value of H?

The strength of the S–Z effect for a particular cluster (the amount by which the temperature of the background radiation has been decreased) tells you how much hot gas there is in the cluster, along the line of sight through the cluster, affecting the photons of the microwave background. But the hot gas in the cluster can, in some cases, also be detected at X-ray wavelengths, from satellites orbiting above the Earth's atmosphere (which, of course, blocks X-rays). The measured amount of X-radiation from the cluster, together with its angular size on the sky, gives a prediction of the strength of the S–Z effect for that cluster, which depends on H because you have to convert the angular diameter into the actual size of the cluster. So measuring the actual S–Z effect and comparing it with this prediction gives you the value of H.

Of course, it isn't quite that simple in practice. Once again, the modelling comes into the calculation. Everything is fine if the cluster is spherical, with the hot gas distributed in a round ball. Then, the depth of the cluster along the line of sight is the same as its width on the sky, and the modelling is straightforward. But if a cluster is actually cigar-shaped, with its long axis pointing towards you, it might look round when in fact the cosmic background radiation has travelled all the length of the cigar, much more than the width of the cluster on the sky. Most clusters probably are roughly spherical, but may be a bit lumpy, or a bit elliptical (or both); the models used can only be based upon a combination of observations and inspired guesswork. In addition, the actual measurements are very difficult, and involve long hours of radio telescope time. As yet, the results from the S–Z effect have very large error bars. But even so, they provide dramatic confirmation that everything I have told you so far in this book is correct.

Even the severest critics of the S–Z technique would tell you, at the beginning of 1998, that it leads to estimates of *H* in the range from 30 to 100 (I know, because I asked those critics for their opinion). They intend this to be a damning indictment of a technique so useless that it cannot even distinguish between the claims of the protagonists in the old debate between the de Vaucouleurs school of thought and the Sandage–Tammann school of thought. But this is missing the point that the S–Z effect is an entirely independent way to measure the Hubble Constant, which can make use of very distant clusters of galaxies and the radiation from the Big Bang itself, cutting out all the steps on the cosmic distance ladder, and not even depending on the calibration of Cepheid variables. Like the gravitational lens technique, it leaps out into the depths of the Universe and gives us a truly cosmic measure of *H* in one bound.

The amazing, wonderful thing is that it gives an answer consistent with the other techniques. In very round numbers, to the nearest power of ten, the value of *H* that you get from the S–Z effect is 50 — not 5, and not 500. To an equal level of approximation, this is the same as the number you get from the traditional ladder based on Cepheids, and from gravitational lensing. The agreement is as important, and dramatic, as the agreement between estimates of the ages of the oldest stars and estimates of the age of the Universe that I discussed earlier. Once again, what matters is not whether one number is 50 percent bigger or smaller than the other, but that they all agree so closely with one another even though they have been obtained by completely different techniques. If the whole theory of the Big Bang was wrong, such an agreement between, in this case, three independent estimates of *H* would be a remarkable coincidence. If you knew nothing about the history of the search for the Hubble Constant, or the ages of stars, but had just invented the S–Z technique, you might have been unsurprised to get a value for H as low as 5 (or 0.5), or as high as 500 (or even 5,000). There is no reason why all three techniques should give answers in the same ballpark unless they are telling us a fundamental truth about the Universe. It was the Bellman, in *The Hunting of the Snark*, who said that "what I tell you three times is true"; the Universe has told us three times, by three completely independent techniques, that the Hubble Constant has a value somewhere around 50. It is true.

But just how close is the Hubble Constant to 50? Critics of the S–Z technique may be unwilling to accept that it sets limits on the value of H better than the range from 30 to 100, but the people who actually use the technique are confident that they can narrow it down rather better than that.

Some of the reluctance to accept the S–Z results at face value stems from the very first published result from the combination of X-ray observations with this technique, back in 1990, which not only gave a surprisingly low value for H (24) but had some exceedingly optimistic error bars ($\pm$ 10). In fact, later studies showed that the modelling used in this work was wrong, and that the number obtained had to be at least doubled. It was, indeed, that kind of result that led to a more thorough investigation of the modelling problems, and more honest assessments of the errors in results published in the mid-1990s. Some typical results from 1995 and 1996, most based on measurements of a single cluster, include an estimate of 38, but including a possible range from 24 to 54, made by a group at the University of Cambridge; 46–91 (with a best estimate of 71) made by a team of Caltech; a value of 54 $\pm$ 14 (from mostly the same Caltech team, but based on studies of three clusters); and 32–67 ("best" value 47), from another study at the University of Cambridge.

While the observers were refining their techniques, the theorists were getting to grips with possible source of error in the model calculations. One study looked at how projection effects (this business of whether a long, thin cluster is pointing towards the observer or not) affect the value of H obtained by the S–Z technique. Another study suggested that in some cases the gravitational lensing of the background radiation by the cluster involved in the S–Z effect might lead to an incorrect estimate of H (I particularly like this example, because it brings home just how complicated any attempt at unravelling cosmological information from the observations is). There are also corrections due to the need to use the theory of relativity to describe properly what is going on in the hot gas in the clusters that produces the X-rays. In addition, it is very well known in the trade that for more distant clusters the measured size of the S–Z effect will depend on the curvature of space—which is actually a good thing, because it means that one day we will be able to use the S–Z effect to work out whether the Universe is open or closed.

But it has to be said that for my present purposes all of these corrections are no more than a shoal of red herrings. The corrections involved are typically a few percent (up to about 10 percent), and they don't all push in the same direction, so they tend to cancel each other out. They really only matter when we think we are measuring the value of H from the S–Z effect to an accuracy better than 10 percent. And we aren't quite there yet, although we are getting close. In 1996, Mark Birkinshaw, of the University of Bristol, carried out a review of all the S–Z data then available, and concluded that the best value for H, based on the data from nine clusters studied by several different teams, was 60, with a scatter of about ± 20.

The best values of the Hubble Constant based on the S–Z effect so far, though, came in during 1997 and early 1998, while I was working on this book, and these take account of at least some of the corrections just mentioned. For three different clusters, a University of Cambridge team found values for H of 38 ($+17$, -12), 47 ($+18$, -12) and 56 ($+18$, -13) — the important point being that all these error bars overlap for a value of H in the low 50s. And what looks to me like the cleanest and best assessment of H from S–Z observations of a single cluster (known as CL 0016+16, this is the third of the clusters studied by the Cambridge team) came from John Hughes, of Rutgers University, and Mark Birkinshaw, in a paper circulated in January 1998. They found a value of 47 ($+23$, -15) for a simple model of the cluster, but after allowing for possible errors due to the geometry and orientation of the cluster, and other factors, came up with a range of 42–61 for H, with a possible further random error of $\pm$ 16 percent. Their best value was worked out for a particular open cosmological model, but the range of possibilities quoted takes account of, among other things, the extreme cosmological possibilities consistent with observations.

With all the improvements made by various teams during 1997, including their own work, Hughes and Birkinshaw then came up with what they call an "ensemble average" value for H, based on the same nine clusters (including CL 0016+16) discussed by Birkinshaw a year earlier, of H = 47.1 $\pm$ 6.8 km/sec per Megaparsec. The precision implied by the decimal points should not be taken too seriously, as they acknowledge; nor should the quoted error bars, although they are getting close to the magic 10 percent. But if you want a single number to quote for the value of the Hubble Constant based on S–Z observations, as of early 1998, that is the one to go for.

All of which seems to have brought us from the controversy of the 1970s to a beautiful consistency — except for two flies in the ointment. But these are only small flies (no more than gnats), and in the second half of the 1990s they began to shrink, suggesting that they may soon disappear (perhaps even before you read these words). The first of these flies was extremely puzzling for nearly two decades, and when I started planning this book my biggest concern was how I was going to explain away the results from the technique known as the Tully–Fisher (or T–F) relation; but a reanalysis of the basis of this technique, using data from the Hubble Space Telescope, has shown that it had been incorrectly calibrated, and brought it almost in line with the results described earlier. The second fly, as we shall see, is the reason why I wrote this book in the first place.

Brent Tully and Richard Fisher didn't exactly invent the technique that now bears their name, but they were the people who, from 1977 onwards, took up the idea, quantified it, and promoted it as a tool with which to measure the value of the Hubble Constant. The basis of the technique is the empirical observation that bigger, brighter spiral galaxies seem to rotate more rapidly than smaller, dimmer spiral galaxies. This is plausible, because a bigger, brighter galaxy is more massive, so it has a stronger gravitational pull, so it can rotate more rapidly without flying apart. But just because it *can* rotate more rapidly doesn't mean that it *must* rotate more rapidly. The T–F relation would only really work as a reliable brightness (and therefore distance) indicator if there were some law of physics which said that a galaxy must always rotate as fast as it can without breaking up. Then, measuring its rotation speed would tell you exactly how massive the galaxy must be, and how bright it must be. As ever, knowing the intrinsic brightness gives you its distance, by measuring its apparent brightness. But there is no such law of nature (or at least, no such law has yet been found) and without one the T–F relation is without any secure foundation in physics — it is, as some astronomers remark disparagingly, "voodoo."

Nevertheless, the voodoo does seem to work, after a fashion. It is relatively easy to measure how fast spiral galaxies are rotating, by studying their emission in the radio band, at a wavelength of 21 centimetres. This radiation comes from hydrogen gas, which is ubiquitous between the stars of a spiral galaxy. In a laboratory on Earth, hydrogen produces a very sharp line in the spectrum at 21 centimetres. Because a spiral galaxy is rotating, however, if it is seen edge on, one side of it is moving towards us, and one side is moving away. So the radiation from some of the hydrogen is slightly blueshifted, to slightly shorter wavelengths, and some is redshifted, to slightly longer wavelengths. The overall effect is that the 21-centimetre line is blurred, and its overall width measured by our radio telescopes tells you how fast the galaxy is rotating.

The snag is that even if the T–F relation between rotation speed and brightness is more than a rough-and-ready approximation (which is doubtful), until the advent of the HST it was hard to calibrate it. There were only half a dozen nearby spirals with accurate distances determined from the Cepheid method, using ground-based telescopes, and only two of these (M31 and M33) are as big and have such large rotational velocities as the more distant galaxies used in applying the T–F technique. There is also a problem with knowing how much to allow for the effect of dust in the distant galaxies, obscuring their own light — since the technique works best for edge-on spirals, where it is easy to measure the velocities, and dust is concentrated in the plane of such a galaxy, this is a big problem. And there is always the problem of Malmquist Bias. On the plus side, the technique can (once it is calibrated as best as possible using local spirals) be applied to very many distant galaxies — Tully and Fisher themselves used a sample of more than a thousand spirals.

Early results from the technique gave high values for *H,* as high as

90, and gave great comfort to advocates of the "short" distance scale favoured by de Vaucouleurs. To put this in a historical perspective, in 1985, when Michael Rowan-Robinson carried out his major review of all the techniques for determining the Hubble Constant, he found that the best value for H was 67 ± 15. This depended on his own somewhat subjective (but informed) assessment of which techniques were the most reliable, and he gave less significance to the more doubtful techniques in working out this number. At that time, the limited supernova data available in the pre-HST days tended to give even lower values for H, and the T–F technique gave the highest values, in the mid-80s. They couldn't both be right. Throwing out the supernova data increased Rowan-Robinson's estimate of H to 78, while throwing out the Tully–Fisher results decreased his value of H to 56 (in each case, with errors roughly ± 15). "This," he said, "gives an indication of the range over which my estimate of H could shift in the immediate future." As we have seen, over the ten years or so following Rowan-Robinson's survey, almost all the results from the other techniques shifted towards the supernova results, and the S–Z effect data (which only became available a decade after this survey) pointed in the same direction. Only the T–F technique was left out on a limb.

By 1996, however, things had begun to change. The technology of ground-based observations had improved significantly (mainly thanks to the use of electronic detectors, CCDs, in place of photographic plates), increasing both the detail of the observations of different galaxies and the number of galaxies with measured Cepheid distances. Riccardo Giovanelli, of Cornell University, used a sample of twelve spirals with known Cepheid distances for a new calibration of the T–F distance scale. Even though this calibrating sample was still uncomfortably dependent on spirals like M31 and M33, which, he acknowledged, are less than ideal calibrators because they show clear signs of distortions in their discs, probably caused by tidal effects, he came up with a value of *H* reduced to 70 ± 5.

But the breakthrough came in 1997, after I had started work on this book (and much to my relief!). One of the Key Projects for the Hubble Space Telescope, a primary reason for its existence, has, from the outset, been to calibrate the distance scale of the Universe by measuring Cepheid distances to as many galaxies as possible, and using these distances, where possible, to tie in the various techniques I have described in this book to a common framework. In 1997, Tom Shanks, of the University of Durham, was able to discard all of the ground-based Cepheid data, and use distances to eleven spirals determined from Cepheid variables by the HST and twelve spirals with distances determined from Type 1a supernovae to make a completely new calibration of the Tully–Fisher relation. Compared with the distance scale determined from the original six ground-based observations of Cepheids in the first calibrating sample of spirals, he found that all the distances determined by the T–F technique had to be revised upwards by nearly 25 percent, and that the estimate of the value of the Hubble Constant had to be reduced accordingly, from 84 ± 10 to 68 ± 8. Among other things, this pushed the distance to the core of the Virgo Cluster estimated by the T–F technique out from 15.6 ± 1.5 Megaparsecs to 19.3 ± 1.9 Mpc, bringing it in line with measurements described earlier in this book.

In an even more comprehensive and statistically powerful reevaluation of the T–F method, involving studies of twenty-four different clusters of galaxies but also using the latest HST data, Giovanelli and his colleagues came to much the same conclusion later in 1997, reporting a best value of the Hubble Constant derived from the technique of 69 ± 5. And then, early in 1998, a team of researchers at the University of Tokyo produced another comprehensive survey, based on a study of 441 spiral galaxies, using the HST Cepheid calibrations, with good statistics and taking account of Malmquist Bias. They came up with much the same value of H as Shanks and Giovanelli, but rather more realistic error bars—a best value of 65, with possible errors of $+20$ and -14. In their own words, there is "a good agreement" between the values of H now being found from the Tully–Fisher technique and those recently obtained by traditional Cepheid-based techniques, supernova studies, and (although they did not specifically mention this) the S–Z effect.

I still have my doubts about the reliability of a technique which produces a shift of 25 percent when the calibration on which it is based is changed from one set of six spirals to another set of eleven spirals, and we surely have not heard the last of changes in the distance scale determined from the Tully–Fisher technique. It is still the weakest technique of any of those described in this chapter. But at least it no longer disagrees with the other estimates; all the error bars overlap. Which leaves us with what looks today like a minor niggle stemming from the Hubble Key Project work, but which in 1994 caused alarm among many astronomers, and eventually led to my own involvement with the search for the Hubble Constant.

8

When Time Began

How We Measured the Age of the Universe

If the consensus described in the previous chapter had emerged as clearly in 1994 or 1995 as it had by the beginning of 1998, I would probably never have got involved with measuring the value of the Hubble Constant, and this book (if it had been written at all) would have ended with Chapter 7. But in the mid-1990s the situation was still a lot more confused than it became just a few years later.

Although my own contribution to the "age of the Universe debate" was by no means definitive, and represents just one minor brick in the scientific edifice, I shall go into some detail about the work I was involved in during the mid-1990s for two reasons. First, it is a chance to describe from the inside a piece of scientific research as it was carried out, false starts, blind alleys, and all. Too often, popular accounts of the scientific endeavour give a distorted version of the truth, in which the march of science seems to be just that — an inexorable forward progression. But that is seldom even close to giving a true vision of what goes on at the cutting edge of research. Second, thanks to the huge public interest in the Hubble Space Telescope, what turned out to be a misleading estimate of the value of the Hubble Constant was given headline treatment in the mid-1990s, and as a result there are still a lot of people who are confused about the best modern understanding of the age of the Universe, and in particular its relationship to the best modern estimates of the ages of the oldest stars.

At that time, several of the new techniques were pointing towards a value of H at the lower end of the range that had been discussed for the previous twenty years, but still not low enough (in most cases) to resolve the discrepancy between the implied age of the Universe and the then best estimates of the ages of the oldest stars. Many astronomers were waiting for the first results from the refurbished Hubble Space Telescope, hoping that Cepheid distances measured with the HST would give a value for H in the 60s, matching the results from the new techniques, if not quite solving the age problem. A few (myself included) were hoping for an even lower value.

We all knew that the HST would be able to tackle the problem, because even before the optics of the telescope had been repaired (in December 1993), it was still operating well enough to be able to resolve Cepheids in the nearby galaxy M81 (just over 3.6 Megaparsecs away). Before the HST was turned on M81, only two Cepheids had been identified in that galaxy, using ground-based telescopes. Even with its flaws, the pre-repair HST was able to identify thirty "new" Cepheids in M81, confirming that everything else on the telescope worked fine, and whetting the appetites of the astronomers for the real stuff. Then, the Hubble Key Project team dropped their bombshell.

You have to appreciate that the Cepheid studies from the refurbished HST didn't come in with a rush — you can't do this kind of work overnight. Even with the advantages this telescope has over its ground-based counterparts, it still has to be turned on to a particular galaxy for a long time. Each observation has to be made twice, at two different wavelengths, in order to make allowance for the effects of cosmic dust, and each single observation takes two orbits of the telescope around the Earth (about eighty minutes) in order to pick out individual Cepheids in that galaxy. After all that, you still have to make such observations from time to time over an intervals of weeks or months in order to measure the periods of those Cepheids. And even the HST cannot pick out objects as faint as Cepheids in galaxies much further away than the Virgo Cluster. By the end of the observing phase of the Key Project, in January 1998, the team had observed Cepheids in three galaxies in the Virgo Cluster, three in another cluster (Fornax), and eighteen so-called field galaxies, not in clusters. The expressed aim of the Key Project was to use data from about twenty galaxies to determine the Hubble Constant to within an error of ± 10 percent, but the final analysis of the data was still under way when this book was being written.

So the Hubble Key Project team (and other people using the telescope) are still dependent on the technique of measuring distances to relatively nearby galaxies and using them to calibrate secondary indicators (such as supernovae, or the properties of galaxies used in the Tully–Fisher technique) and treating them as stepping-stones to move further out into the Universe at large. As the team themselves have always acknowledged, the resulting dependence on the Virgo Cluster introduces a possible error of ± 20 percent right at the beginning of the calculation, for all the reasons that I have described earlier. But even allowing for all this, the reaction of most astronomers was stunned amazement when the Hubble Key Project team announced, in October 1994, that their first observations of twelve Cepheids in a Virgo Cluster galaxy (M100) led to a value for *H* of 80 ± 17.

This headline figure for the first determination of the Hubble Constant using the Hubble Space Telescope did just that — it made headlines, and revived the saga of the conflict between the ages of the oldest stars and the implied age of the Universe. But the authors of the articles under those headlines failed to take note of the caveats buried deeper in the report. The Key Project members acknowledged that the distance to M100 (which had, after all, been chosen because it was relatively easy to study) might not be the same as the distance to the core of the Virgo Cluster, and this could produce a large error in their calculation. They also chose a particular value for the infall velocity to Virgo, corresponding to a cosmological recession velocity of 1,400 kilometres per second, but pointed out that choosing a recession velocity of 1,180 km/sec (favoured by other astronomers) would reduce their estimate of *H* to 69 ± 14, for a Virgo distance (from their studies of M100) of 17 Megaparsecs. And many astronomers argue that the true distance to the Virgo core is more than 20 Mpc (so M100 is very much on our side of the cluster), which would have the effect of reducing the estimate of *H* based on secondary stepping-stones still further.

The rough-and-ready way to measure *H* from Virgo Cluster data, remember, is not to use just the redshift of any individual galaxy, such as M100, which will be affected by its own random motion, but to use the core distance and the average redshift of very many galaxies in the cluster, in effect placing them all at the distance of the core and trusting that their random velocities will average out to zero. So changing your estimate of the distance to the core by 20 percent changes your estimate of *H* by 20 percent.

In fact, the main reason why this first Virgo Cepheid result from the Hubble Key Project team came up with such a high value for H was that the analysis was contaminated by exactly the same kind of astronomical optimism that had contaminated the work of de Vaucouleurs for so long, and by the choice of parameters for the key properties of the Virgo Cluster. This very first result from the Key Project was actually entirely consistent with a value for H of 55 — but that wasn't the message that came across, although in a more technical paper published in August 1995, even though the team stuck by their estimate of $H = 80 \pm 17$, they also expressed the result in rigorous statistical terms, using standard ways of measuring probability, by saying that it was 95 percent certain that the value of the Hubble Constant lay in the range from 50 to 100. Turning this around, that means that there is no more than a one in twenty chance that H is actually either smaller than 50 or bigger than 100, and a nineteen in twenty chance that it is in the range 50–100 — which is rather a different story from the one portrayed in the headlines triggered by the 1994 paper.

Since 1994, as Cepheid data from more galaxies have slowly been gathered in by the HST, the estimates of H published by the Key Project team have slowly come down, but without making headlines. The work is so time-consuming that each "new" galaxy studied justifies a new scientific paper, but by the summer of 1997 in a summary of the state of play the Key Project team quoted a best value for H of 73, with total estimated error bars of ± 14. But that is getting ahead of my story.

All that had really happened was that in the mid-1990s the old debate between proponents of the "long" and "short" distance scales had moved out into space — a point forcefully brought home when, early in 1996, Sandage, Tammann and five of their colleagues used

HST Cepheid data for the distances to nine galaxies to calibrate both the Tully–Fisher relation and the supernova distance scale. They came up with a distance to the Virgo core of 22 Mpc, and used a cosmological recession velocity for the cluster of 1,178 km/sec, plus the supernova data, to come up with a value for H of 55 ± 10. "Systematic errors," they said, "tend to make this an upper limit; in particular, the case H bigger than 70 can be excluded."

Two different teams, using data from (in many cases) the same galaxies, obtained with the same telescope, had come up with two estimates of H that barely agreed with one another even at the extreme range of their published error bars — 55 + 10 is 65, while 80 − 17 is 63. The naive resolution to the puzzle might have been to split the difference and call it 64 (pleasing neither camp). But the very name Hubble Key Project seemed to give the results from that team an aura of mystique that made them more credible.

Ever since the Key Project results for M100 had first been published, late in 1994, I had bored my colleagues at Sussex University by pointing out, when the work was debated, that they couldn't possibly be right, because that would imply that our Milky Way Galaxy was a huge and unusual spiral galaxy. Nobody took much notice, except to say that "you can't generalise from a sample of one," meaning that for all we know our Galaxy *is* unusual — without having more information about galaxy sizes in general, how can we tell? But when the Sandage team reported their analysis in March 1996, I gleefully pointed out this new ammunition in support of my case, only to receive the reaction, "Well, Sandage would say that, wouldn't he?" But by then, as the paper from Sandage's team highlighted, there were more than a handful of HST Cepheid distances to spiral galaxies. Enough, it turned out, to offer a completely new way to measure H.

It was a couple of months after the Sandage and Tammann interpretation of the HST data appeared that the penny dropped. There was a seminar at Sussex University in which the speaker described some of the evidence, from observations of Cepheids in the Virgo cluster, in favour of the short distance scale and high value of the Hubble Constant. For once, I kept quiet during the talk. But afterwards, walking down the corridor with a group of students, I took the opportunity to make my familiar case for the long distance scale to a relatively new audience, pointing out that everything we had just been told implied that the Milky Way was an unusually large galaxy. "Well," said one of them, "we could find out."

"What do you mean?"

"With all the HST data, and the ground-based stuff, there must be plenty of Cepheid distances to spirals by now. We could find out all the sizes, and compare them with the Galaxy."

It was a classic example of not being able to see the trees for the woods. I had been so focused on the relationship between galaxy sizes and the value of the Hubble Constant that I had overlooked the fact that it might be possible to just measure the sizes of spirals in our neighbourhood. All we had to do was take Cepheid distances to all the nearby spirals, and compare them to the apparent angular diameters of those same spirals to work out the true linear diameters, in kiloparsecs, by triangulation. Then, we could find out definitively if the Milky Way was an average size or not. And if it *did* turn out to be average (or even if it didn't), that would immediately tell us something about the value of H!

It was such a good idea that it seemed strange nobody had thought of it before — but even by combining the available Cepheid data from ground-based telescopes and the new HST data there had only (in the summer of 1996) been enough known distances to spirals to make such an assessment worthwhile for a few months. Simon Goodwin, the student who had had the bright idea, agreed that it would be worth pursuing, provided nobody else had done it already. A computer search of the astronomical databases showed no trace of a scientific paper preempting such an investigation, but Simon was in the midst of writing up his doctoral thesis, and was understandably reluctant to drop everything and get stuck into a new project immediately. Without his familiarity with the databases and the latest computer techniques, by the time I had made any progress on my own he would have finished his thesis anyway, so the only thing to do was be patient. As far as Simon was concerned, there was no real rush; he was clearly happy to humour a member of the older generation, but didn't see any great urgency in the project.

Things drifted a little in the summer, but Simon's enthusiasm for the idea soared when the third member of the team came on board. Martin Hendry, although not much older than Simon, had already developed a formidable reputation in astronomy, and was particularly knowledgeable about the state of play in the various attempts being made to measure the Hubble Constant. He is also a cautious Scot, and an expert in the statistical techniques astronomers need to use in order to assess the significance of their data (indeed, he invented some of those techniques). Simon suggested that we ask Martin if the project really was worth pursuing, and if he would like to come on board. Martin enthusiastically agreed that it would work, and that a new, independent way to measure H was just what was needed to cut through the confusion that still existed about the cosmic distance scale and the age of the Universe. And this endorsement from one of the leading lights of his own generation fired Simon's enthusiasm, encouraging him to move straight on to the spiral data as soon as he had finished writing up his thesis.

It turned out that there were just enough spirals about which just enough information was known for us to determine, once and for all, the relative size of the Milky Way. Taking all the available Cepheid data from ground-based telescopes and the HST, and considering only those spiral galaxies which have a close physical resemblance to the Milky Way (in terms of how tightly the spiral arms are wound, and so on) we had just seventeen galaxies to work with, all with accurately determined distances. As ever in astronomy, things are not as simple in practice as they look when a project is planned, and the key problem in this case was deciding just how you measure the edge of the image of a spiral galaxy on a photographic plate or CCD image, and how you compare this with an equivalent measurement of the edge of the Milky Way. Fortunately, there is a standard way to measure galaxy angular diameters, in terms of the way the brightness of the galaxy drops off moving outward from its centre. In effect, you draw a contour line around the galaxy image at the position where the brightness drops to a certain level, and call that the edge of the galaxy. This defines a distance known as the isophotal diameter, and all we had to do was look up the isophotal diameters of our chosen galaxies in the catalogues.

After all the excitement, though, it turned out that the hardest thing was to determine the size of the Milky Way itself in the equivalent fashion. Since we sit inside the Milky Way Galaxy, we have to use different techniques for measuring its size, and astronomers are still arguing about the exact diameter of the Milky Way (or even what you mean by the expression "the exact diameter of the Milky Way"). But all we were concerned about was the equivalent isophotal diameter of the Milky Way, which can be calculated from the observed distribution of stars within the Milky Way, so that we can work out what the Galaxy would look like from outside. This gives us a calculated isophotal diameter for the Milky Way of 26.8 ± 1.1 kiloparsecs.

How do we know that our formula for the isophotal diameter of the Milky Way is correct? Fortunately, there are two spiral galaxies close enough for us to be able to measure the distribution of stars in them and use the same formula to work out a theoretical isophotal diameter which we can compare with the observed isophotal diameter to see how good the formula is. For our old friend M31, the formula was out by 5 percent, and for M100 it was out by just 0.3 percent. Compared with the traditional uncertainties in the value of H (and most other astronomical measurements), this is OK.

When we had done this, it was straightforward to use the distances and isophotal diameters of the seventeen spirals in our sample to work out the actual linear diameter of each of them, from simple geometry, and take the average. It came out as 28.3 kilopars, marginally (but not significantly) bigger than the size of the Milky Way. To my delight, we had found that the Milky Way is indeed an average spiral. And when we reduced our sample slightly, including only the twelve galaxies that most closely resemble the Milky Way in appearance, the average went up to 33.6 kpc. With such small numbers, we shouldn't read too much into this, since a diameter of 26.8 kpc for the Milky Way is still barely significantly smaller than the average.

The caveats are important, because, thanks to Martin Hendry, in spite of the small numbers involved our work was based on a sound grounding of statistical analysis, which confirmed that all the galaxies in our sample were members of the same family of objects (the same statistical population) and that it was meaningful to work out an average diameter in this way.

My delight at the result was tempered just a little, though, by discovering that my critics had been right all along about one thing — you can't generalise from a sample of one. In a paper published in 1993, Allan Sandage had done just that. He had picked a galaxy

(M101) that looks like a "typical" spiral, and which has a known Cepheid distance. Then, by assuming that the linear diameter of M101 is exactly the average for spirals, he had used the kind of argument I used to use for the Milky Way to claim that the best value for the Hubble Constant was 43 ± 11, on the assumption "that M101 is not the largest in a distance limited sample." Our analysis showed that — you guessed — M101 is indeed the largest galaxy in our neighbourhood (which is what is meant by a "distance limited sample"), with a whopping diameter of 61.8 kpc.

Our results immediately suggested, of course, that the actual value of *H* must be "about" 50. But the salutary lesson of M101 meant that even I agreed on the need to carry out a thorough statistical study of as many galaxies as possible before rushing into print. The work on the size of the Milky Way, though, stood on its own. It was completed by Christmas 1996, and we sent it off for publication* before turning our attention to the thousands of spirals with known redshifts and angular diameters in the standard catalogues. And while we did so, news came in from the Hipparcos satellite that had a direct bearing on our search for the age of the Universe.

Hipparcos, a European Space Agency project, had been launched on 8 August 1989, but never reached its intended high geostationary orbit. Instead, because of the failure of the rocket motor meant to boost it into this orbit, it stayed in a highly elliptical orbit which took it swinging from a high of 35,000 kilometres above the Earth to a low of 500 kilometres above the Earth. Apart from anything else, this meant that the spacecraft passed through the Van Allen radiation belts that surround the Earth twice in each orbit, with its solar panels and electronic equipment taking a hammering from the particles in the

---

* It wasn't actually published until 1998, but that was the fault of a referee and an editor who didn't understand why the work was so important!

radiation belts each time. It looked at first as if the mission might be a complete failure. But the astronomers and engineers running the project managed to find ways to work around the problems, keeping the satellite working for four years (a year longer than its planned lifetime), during which it measured the parallaxes of nearly 120,000 stars to an accuracy of 0.002 arc seconds, using its modest 29-centimetre-diameter mirror telescope (Hipparcos did not need a large telescope because it did not have to study very faint objects; its key asset was being able to measure parallaxes, and therefore distances, to unprecedented accuracy from above the blurring effects of the Earth's atmosphere).

Hipparcos sent back more than 1,000 gigabytes of data to the astronomers back on Earth. But they had to exercise enormous patience in waiting for the results from the analysis of the huge data set, because the way it was generated meant that it had to be processed as one block of measurements. The astronomers could not get a single measurement out of the processing until all the data had been processed together, and then they got all the measurements at once. The processing took nearly as long as it had taken to make the observations in the first place, which is why the results from Hipparcos were not released until 1997, when we were in the midst of our work on the Hubble Constant. The result was a map of the stars in three dimensions, with their positions on the sky defined to an accuracy of one thousandth of a second of arc, described by one member of the team as being equivalent to being able to pick out a golf ball on the top of the Empire State Building using a telescope on the top of the Eiffel Tower.

It's a sign of how difficult all previous attempts to establish the baseline for the cosmic distance ladder had been that Hipparcos provided the *first* direct measurements, by parallax, of the distances to Cepheids. Before, the few measured Cepheid distances, as I described in Chapter 3, rested on more indirect, statistical methods of measuring the distances to a few key stars — just eighteen Cepheids. But Hipparcos provided directly measured distances not just to a few Cepheids, but to 220 of these stars, and detailed analysis based on the twenty-six Cepheids with the most accurately determined parallax distances provided a definitive calibration of the Cepheid distance scale (incidentally, that small proportion of Cepheids, 220 out of 120,000 stars studied, accurately indicates how rare Cepheids are, which is another reason why previous generations of astronomers had so little data to work with).

It turned out that Cepheids are slightly brighter, and slightly further away, than had previously been thought, increasing the cosmic distance scale (and the implied age of the Universe) by about 10 percent. But at about the same time that this work was being announced, another study of the way the Cepheids in the Large Magellanic Cloud had been used as a stepping-stone to more distant galaxies suggested that the standard Cepheid distance scale might have to be reduced, probably by about 5 percent. By and large, the two corrections seem to cancel each other out; but it is worth bearing in mind that one implication of the Hipparcos survey is that all the values for H quoted in the previous chapter and in this one might have to be reduced slightly (and the estimated age of the Universe increased in proportion), though not by as much as 10 percent.

The really exciting aspect of the Hipparcos survey, though, was its implications for the ages of globular clusters in our Galaxy. By recalibrating the brightnesses of the appropriate main sequence stars, using distances to nearby stars determined accurately by parallax, astronomers working with Hipparcos data found that the globular clusters are significantly further away than had previously been thought. This result, announced early in 1997, means that the stars in those clusters must be brighter than used to be thought, in order to look as bright as they do on the sky. And because an intrinsically brighter star burns its nuclear fuel more quickly than an intrinsically fainter star, this means that the oldest stars in the Galaxy are younger than used to be thought — a younger, hotter star (or cluster) mimics the appearance of an older, fainter star (or cluster) because it has processed its nuclear fuel more rapidly.

Hipparcos reduced the best estimate of the age of the oldest known globular clusters from sixteen billion years to eleven billion years, at a stroke making life that much easier for cosmologists trying to fit their estimates for the age of the Universe (the time since the Big Bang) around the estimate of the ages of the oldest stars. To be precise, allowing for remaining uncertainties the best estimates for the ages of those oldest stars came down from the range of sixteen to eighteen billion years to the range of eleven to thirteen billion years. And three different teams, between them using Hipparcos data in two different ways on three different sets of stars, came up with exactly the same result. Along the way, incidentally, this work also gave a new estimate of the distance of the Solar System from the centre of the Milky Way, as 8.5 ± 0.5 kiloparsecs.

There were (and still are) some puzzling aspects of the Hipparcos survey. For a few nearby systems (notably the Pleiades open cluster, the "Seven Sisters" in the constellation Taurus), the distances implied by Hipparcos change the estimates of the ages of the stars by more than can be comfortably accommodated by standard models of stellar evolution. But it is the job of theorists to come up with models that match the observations, not the job of observers to bend their data to match the models, and in any case the debate stirred by these studies of open clusters has little or nothing to do with the story of the ages of globular clusters.

The definitive study of all the evidence available on globular clusters to date appeared in the summer of 1997, just when we were finishing our work on the Hubble Constant. It came from a group headed by Brian Chaboyer, of the University of Arizona, and it used not only the Hipparcos parallaxes but four other independent techniques to come up with the best estimate yet for those ages. It turned out that the Hipparcos data were, in fact, just one straw in a wind that had already been gaining strength, and that with the continuing improvements in ground-based telescopes, electronic detectors, and computers to handle the statistics and run the stellar models, even without Hipparcos by the mid-1990s an upward revision in the globular cluster distances and a downward revision of the globular cluster ages was required. What mattered was that all of the evidence pointed the same way—and what mattered most of all, making this a defining moment in such studies, was that Hipparcos agreed with all the ground-based techniques.

Chaboyer and his colleagues came up with a best estimate for the ages of the oldest globular clusters in our Galaxy of 11.5 ± 1.3 billion years, with only a one in twenty chance that the ages could be below 9.5 billion years. As they pointed out, even the highest value of the implied lower limit on the age of the Universe could be matched up with the time since the Big Bang even in a flat, Einstein–de Sitter model of the Universe, if the value of the Hubble Constant is less than 67 km/sec per Megaparsec. And that, as we already knew when we read their paper in June 1997, was no problem at all.

It had taken us so long to come up with our own definitive value for H because we were determined (doubly determined, after seeing what had happened when Sandage tried to put all his eggs in the M101 basket) to base that value on a really thorough statistical analysis of the best available data. That still wouldn't have taken six months;

but unfortunately (for myself and Simon Goodwin — it was a step up the academic ladder for him) the statistical wizard in the team, Martin Hendry, moved to a more senior job at Glasgow University, and wasn't able to devote much time to our project for several weeks.

Even without the best statistics, it was easy to see which way the wind was blowing. In his analysis based on the size of M101 as a "typical" spiral, Sandage had used just eighty-six field galaxies for investigation, estimating their distances by comparing their apparent angular diameters with the apparent angular diameter of M101, and then comparing these distances with their redshifts to estimate H. Just rescaling Sandage's data to the average spiral diameter that we had determined, instead of to the diameter of M101, but using the same eighty-six field galaxies, increases this estimate of H from 43 to 57 (this kind of quick'n'easy approach appealed to me, but didn't cut much ice with my colleagues). But we wanted a lot more galaxies to work with, and settled on a catalogue known as the Third Reference Catalogue of Bright Galaxies, or RC3, for our study (somewhat iron-ically, in view of our conclusions, this catalogue was based on observa-tional work by a team headed by Gérard de Vaucouleurs; but he always was a fine observer, and we chose it because it was the best of its kind). This gave us 3,827 spirals to play with, all with known redshifts (the largest equivalent to a recession velocity around 20,000 km/sec) and whose angular diameters had been measured, in accordance with the rules described earlier. This was where the statistics came in.

First, we tidied up our nearby sample and our estimate of the average size of a galaxy like the Milky Way. Originally, we had had seventeen other galaxies plus the Milky Way itself, making a total of eighteen spirals on which to base the average — as it happens, exactly the same number as the number of Cepheids with known distances on which the cosmic distance scale was originally based (see Chapter 3).

But in order to be scrupulously honest, we eliminated six galaxies from the sample, leaving us with the twelve (including the Milky Way itself) most like the Milky Way in overall appearance — it is a key feature of our analysis that we were trying to compare like with like. The different kinds of spiral galaxy, classified mainly by how open or how tightly wound the spiral pattern is, were originally defined by Hubble himself, in the 1920s, and in terms of this classification we used only galaxies with "Hubble Types" between 2 and 6.

What we actually found when comparing the average size of this local sample of galaxies with the average size of the galaxies in the RC3 catalogue not only gave us a value for H, but also highlighted in miniature the reasons why there had ever been two schools of thought about the value of the Hubble Constant. The way we make the comparison is simple. We have thousands of galaxies with known apparent angular diameters (the isophotal diameters, measured in the same way as for our local sample) and known redshifts, but we don't (yet) know the value of H. Each angular diameter converts, by triangulation, into a true linear diameter when multiplied by a number which depends on the redshift (known) and H (not known), which together give the distance to each galaxy. So for any chosen value of H, we can work out the appropriate "true" linear diameter for each galaxy in the sample, and take the average for the whole sample. All we have to do is find the single value of H that makes this average the same as the average linear diameter for the local galaxies with Cepheid distances. It would have been tedious, but not difficult, for Hubble and Humason to have done this, using pencil and paper, if they had had the benefit of the RC3 catalogue in 1930; for us, with a computer to do the donkey work, it was no effort at all. But we still had to be careful about comparing like with like.

If we behaved with blind optimism and took the RC3 catalogue at face value, assuming that there were no observational biases in the sample (that is, biases introduced by the mechanics of the observing process, not by any prejudice of the observers themselves), and took the average in this way, it gave a very nice match to the average of our local sample for a value of H of 80. But we know this cannot be right, because we know that the sample is biased by the limitations of our telescopes.

The big problem as ever, is Malmquist Bias (in this case, a kind of geometrical Malmquist Bias). The RC3 catalogue only includes galaxies which have an angular diameter on the sky bigger than 1 minute of arc. So it must be biased towards large galaxies, since small galaxies at large distances just won't be picked up in the catalogue. The Milky Way itself, for example, would not show up in this catalogue at a distance corresponding to a redshift of more than about 5,000 km/sec. So the first thing to do was to throw out of the sample all the galaxies with very high redshifts, where this problem is most acute. We also threw out all the galaxies with very low redshifts, because for such nearby galaxies the local random motions of the galaxies will produce Doppler effects comparable to the cosmological redshifts we are trying to use, confusing the issue. So in the detailed calculation we used a subset of the RC3 galaxies with redshifts between 1,500 and 5,000 km/sec (in round terms, out to the point where a galaxy the same size as the Milky Way would still just be detectable). For a value of H of 50, that corresponds to a distance of 100 Megaparsecs, about five times the distance from here to the Virgo Cluster core. This still gave our statistician 1,388 galaxies to work with (all with Hubble types in the range 2–6), enough for him to apply a battery of statistical techniques confirming that what we had was what the statisticians call a well-behaved sample obeying Gaussian statistics (named after the great mathematician Karl Gauss).

We still weren't quite out of the woods. An optimist who assumed that this was all we had to do to the catalogue sample would have compared the average sizes of the galaxies in the reduced sample with the average for our local calibrating sample, and come up with a value for H of 60. But we could do better than this, by allowing for the residual effect of geometrical Malmquist Bias in the sample (essentially, by looking at how the number of small galaxies seems to drop off as you look further out into the Universe, and allowing for the missing galaxies in a proper statistical way). When our analysis was complete, we came up with a best estimate of H of 52 ± 6 km/sec per Megaparsec. We had measured the value of the Hubble Constant. And I have to admit to a certain private feeling of glee that we had done this almost to the 10 percent accuracy that was the aim of the Hubble Key Project, partly using their Cepheid distances, and before they had reached a similar accuracy themselves.

Applying a battery of further statistical tests, Martin Hendry also showed that there is only a one in twenty chance that H could be bigger than 75 (and, correspondingly, just a one in twenty chance that it could be less than 35). Slightly less cause for glee there, perhaps, but it did tie our work firmly in to the consensus described in Chapter 7 — and even to the Hubble Key Project itself, where the equivalent "95 percent confidence" limit quoted by the team, as of the beginning of 1998, was from 55 to 85, in the usual units.

The agreement between our result and the result from the Hubble Key Project team may be even better than it looks at first sight, since one thing I haven't mentioned is that it is almost certain that the galaxies picked out by the Key Project team to search for Cepheids are slightly larger than the average spiral, simply because of the natural tendency to go for the biggest, brightest objects they could see. If so, our whole result should be shifted bodily (error bars and all) in the direction of larger values of H. But not too far — in 1999, just as this edition of the book was going to press, Martin Hendry and Stephane Rauzy carried out a new analysis which tried to take account of this effect, and suggested that the corrected galaxy diameter technique might yield a value of H in the mid-60s. At the same time, the Key Project team announced their final value for H, which, allowing for errors, covered the range from 65 to 77, in the usual units. I'm still suspicious that this result is a little too high, for the reasons discussed earlier, and my own "best guess" is that H is about 60. But what remains to be fixed up is largely a matter of detail, so this is a natural time to mark the end of the first phase of the investigation of the age of the Universe.

And don't neglect the broader perspective. As well as measuring the value of the Hubble Constant, we had found that the Milky Way is slightly smaller than the average spiral, and that there are some giant spirals out there four times bigger than the Milky Way. It had taken sixty-four years for astronomers to establish that what Eddington had regarded as common sense was correct. Apart from the exact value of H that we came up with, the important implication of finding this particular value of H is that we do indeed live in a typical part of the Universe. The principle of terrestrial mediocrity is valid, so it is reasonable to draw general conclusions from our investigations of the galaxies available in our neck of the woods — it is, indeed, worth doing cosmology.

Although, by the time we had completed our work (but not when we started), there was an impressive agreement between many different techniques that H does indeed lie in the range from 50 to 60, we like to think that there is something special about our approach. First of all, of course, it is based on Cepheid distances, and everyone agrees that Cepheids are the best and least controversial distance indicators, the most securely based step on the traditional cosmological distance ladder, with errors of at most a few percent. But then, in order to move further out into the Universe than Cepheids alone can take us, we used a completely geometrical technique, as sound as parallax itself, the opposite of the "voodoo" approach. As Virginia Trimble (among others) has pointed out, only geometrical methods are actually *measuring* distances across the Universe; all the other techniques are "mere indicators," requiring some additional steps of reasoning or inference, and at least some input from physics. In order to understand our method of measuring *H,* you don't need to understand anything except the geometrical surveying techniques used every day here on Earth. Which just leaves us with the slightly more tricky task of converting our securely based *measurement* of the Hubble Constant into an age of the Universe.

If the Universe had been expanding at the same steady rate ever since the Big Bang (if the Hubble Constant really were a constant, and not a parameter that decreases as the Universe ages), a value for *H* of 50 km/sec per Megaparsec would mean that the Universe is twenty billion years old, ample to accommodate the ages of the oldest known objects in the Universe. Indeed, this age would be *too* ample, since it would imply that nothing at all happened to leave a mark on the Universe today for the first five billion years or more of its existence—25 percent of its lifetime to date. But, as I explained earlier, the effect of gravity is to slow the expansion down, so that *H* was bigger in the past than it is today. The Universe is correspondingly younger than you would infer without allowing for gravity at work in this way, because it was expanding faster long ago, and took less time to reach its present state than it would have if it had always expanded at the rate we see today.

The question is, how much does this reduce our estimate of the age of the Universe? In the simplest Einstein–de Sitter model, for a flat Universe with exactly the critical density of matter, the age has to be reduced by about 30 percent, to a little over thirteen billion years. Many cosmologists would like this to be the correct description of the Universe, because it is so simple, and because their favoured model of the Big Bang, inflation, says that the Universe should be very nearly flat (see my book *In Search of the Big Bang*). But the observations cannot yet prove this, and all they can tell us is that there must be at least a third of the critical density of matter in the Universe as a whole. With only a third of the critical density, the Universe would be just open, not flat, and would not have slowed down so much since the Big Bang, so the value of *H* measured today would be a more accurate guide to the overall rate at which the Universe had been expanding during its lifetime. In fact, the reduction factor is now about 20 percent, instead of 30 percent. So we have to reduce the implied age of the Universe from twenty billion years to sixteen billion years.

Without invoking a cosmological constant (if you do, you can have any age you like by choosing the appropriate value for the constant), this gives us the range of possible ages for the Universe, if the Hubble Constant has a value of 50. It is at least thirteen billion years old, and may be as much as sixteen billion years old. And remember that, thanks partly to Hipparcos and also to the other studies collated by Chaboyer and his colleagues, our estimates of the ages of the oldest known things in the Universe, the globular clusters, now lie in the range from ten to thirteen billion years, with a best value of 11.5 billion years. Notice that the two sets of numbers no longer overlap. For the first time in the history of astronomy, in 1997 the two numbers, for the ages of the oldest objects in the Universe and the age of the Universe itself, were on the right side of each other. Even allowing a billion years for the first star systems to form after the Big Bang (about right, according to the computer models), and even with the simplest model of the Universe, which gives the youngest possible age, there is no conflict. Far from it — there is the most impressive agreement between two numbers derived in completely independent ways, using totally unrelated techniques to investigate totally different phenomena — the expansion of the Universe and the evolution of stars.

It is hard to overstate the importance of this discovery. Unless we are the victims of the most cruel coincidence, it means both that we really do understand the way stars behave, and that the Universe really did begin at a finite moment in time — or rather, that there was a beginning to time itself. Seventy years ago, a single human lifetime, the workings of the stars were still largely a mystery, and the idea of a birth of time had scarcely been aired in scientific circles. To any previous generation, it would have smacked of magic to have claimed to understand the inner workings of the stars *or* the birth of time. Now we understand both, and both pieces of magic tell us the same story about the age of the Universe.

In a previous book,* I talked about the way particle physicists probing the inner structure of matter hope to find a deep truth about the way the forces and particles of nature work, combining their description in one mathematical package, the theory of everything. But that, to most of us, is a rather abstract and esoteric idea. Nobody has seen a quark. By contrast, though, all of us have seen the stars, and few can have failed to wonder about their significance; science probably began when our ancestors first turned their eyes to the heavens

* *In Search of SUSY*.

and began to wonder what the stars are and how they got to be there. If anything in science really is a Deep Truth, it is the match between stellar astrophysics and cosmology, and the discovery of the birth of time. I'm amazed that I have been lucky enough to play a small part in establishing this truth about the Universe, and delighted to be able to share it with you.

Afterword

The Big Picture

During 1999, several new pieces of evidence emerged which together confirmed that the age of the Universe is a little over thirteen billion years, but which give a slightly different perspective on the big picture of cosmology. The story that got the biggest splash in the media was the discovery, from studies of very distant supernovae, of evidence suggesting that the expansion of the Universe is actually getting faster, not slowing down, as the Universe ages. The effect can be expressed mathematically in terms of Einstein's cosmological constant, lambda; but in physical terms it can be understood in terms of an energy of the vacuum, an energy possessed by empty space. One effect of this energy is to make space springy, so that it expands. But remember that mass and energy are related by Einstein's equation $E = mc^2$ — so the presence of the vacuum energy increases the mass (and therefore the density) of the Universe. That provides an extra gravitational pull, which tends to slow down the expansion. Throughout most of the history of the Universe so far, the effects have been very nearly in balance, so the cosmological constant has not had a profound effect on how the Universe got to be the way it is today. This means that the story I have told in this book, about how astronomers have measured the age of the Universe, is essentially unaffected by the discovery that there is a small cosmological constant. But in so far as here is an effect on the calculations described here, it works in the "right" way. If the expansion of the Universe has got a little bit faster because of the cosmological constant, that means it was expanding a little bit more slowly in the past and took longer to get from the Big Bang to its present state. In other words, you can have slightly higher values of the Hubble parameter, H, today (in the sixties rather than in the fifties), and still have an age of the Universe above thirteen billion years. In the future, as the Universe expands further and its matter density goes down, the vacuum energy will come to dominate, and the acceleration will begun to run away, with profound effects on the ultimate fate of the Universe — but that is another story.

The supernova evidence suggests that the amount of energy stored in the vacuum in this way is about enough to provide 70 percent of the critical density needed to make the Universe flat, in the way Einstein and de Sitter hoped it might be. Because we know from other observations that there is about 30 percent of the critical density around in the form of matter, these studies immediately suggested that Einstein and de Sitter were right. The case for a flat Universe became incontrovertible a little later in 1999, thanks to new observations of the cosmic microwave background radiation, made with instruments carried aloft on balloons. These are very subtle observations of the details of the pattern made by the radiation on the sky (the way its temperature varies minutely from place to place on the sky), which is influenced by the shape of the space through which the radiation has traveled. They show that the Universe really is flat, in the Einsteinian sense. Because we can see evidence for only 30 percent of the critical density in the form of matter, these observations are also telling us, independently of the supernova studies, that 70 percent of the critical density is provided by vacuum energy, the lambda term in Einstein's equations.

The effect of all this on astronomical calculations of the age of the Universe was summarised, at the end of 1999, by Charles Lineweaver, of the University of New South Wales. Pulling all of the evidence together (including some subtle phenomena I have not discussed here), Lineweaver indeed came up with a value for H in the sixties (consistent with our final value), and an age for the Universe of 13.4 ± 1.6 billion years. (The subtleties, as well as the lambda term, slightly alter the simple relationship between H and the age of the Universe that I used earlier.) This is the best estimate yet of the age of the Universe, and I am happy to say that the slightly less accurate age determination described in this book, which gave a best value of a little over thirteen billion years, fits in beautifully with Lineweaver's number.

Further Reading

The following books will enable you to find out even more about the age of the Universe debate, and the discovery of cosmic time. All are fairly accessible, but the ones marked with an asterisk might be a little more intimidating to anyone with an aversion to equations.

Gale Christianson, *Edwin Hubble* (Farrar, Straus and Giroux, 1995)

Stuart Clark, *Towards the Edge of the Universe* (Wiley, 1997)

*Peter Coles and George Ellis, *Is the Universe Open or Closed?* (Cambridge University Press, 1997)

Arthur Eddington, *The Expanding Universe* (Cambridge Science Classics, 1997; first published by Cambridge University Press, 1933)

John Gribbin, *Companion to the Cosmos* (Orion, 1996)

John Gribbin, *In Search of the Big Bang* (revised edition, Penguin, 1998)

Edwin Hubble, *The Realm of the Nebulae* (Dover edition, 1958; first published by Yale University Press, 1936)

Helge Kragh, *Cosmology and Controversy* (Princeton University Press, 1996)

Alan Lightman and Roberta Brawer, *Origins: The Lives and Worlds of Modern Cosmologists* (Harvard University Press, 1990)

Malcolm Longair, *Our Evolving Universe* (Cambridge University Press, 1996)

Dennis Overbye, *Lonely Hearts of the Cosmos* (HarperCollins, 1991)

*Michael Rowan-Robinson, *The Cosmological Distance Ladder* (W. H. Freeman, 1985)

Harlow Shapley, *Galaxies* (third edition, revised by Paul Hodge, Harvard University Press, 1972; originally published, 1943)

*Robert Smith, *The Expanding Universe* (Cambridge University Press, 1982)

Index